T0133307

# Voices for the Watershed

Great blue heron. Photo: Frozenrope

# Voices for the Watershed

## Environmental Issues in the Great Lakes–St Lawrence Drainage Basin

Edited by Gregor Gilpin Beck and Bruce Litteljohn
for the Wildlands League chapter of
the Canadian Parks and Wilderness Society

McGill-Queen's University Press
Montreal & Kingston · London · Ithaca

ISBN 0-7735-2003-1
Legal deposit second quarter 2000
Bibliothèque nationale du Quebec

Printed in Canada on acid-free paper

McGill-Queen's University Press acknowledges the financial support of the Government of Canada through the Book Publishing Industry Development Program (BPIDP) for its publishing activities. It also acknowledges the support of the Canada Council for the Arts for its publishing program.

**Canadian Cataloguing in Publication Data**

Main entry under title:
Voices for the watershed : environmental issues in the Great Lakes–St Lawrence drainage basin

Includes bibliographical references and index.
ISBN 0-7735-2003-1

1. Water–Pollution–Saint Lawrence River Watershed. 2. Water quality management–Saint Lawrence River Watershed. 3. Water–Pollution–Great Lakes. 4. Water quality management–Great Lakes. 5. Environmental protection–Saint Lawrence River Watershed. 6. Environmental protection–Great Lakes. I. Beck, Gregor Gilpin, 1964- II. Litteljohn, Bruce III. Title.

GB992.C3V65 2000 363.739'4'09714 C00-900066-6

Preparation of this book was funded in part by a grant from the National Oceanic and Atmospheric Administration's (NOAA) National Undersea Research Center, at the University of Connecticut Award No. NA46RU0146, subgrant UCAP-94-21, and the Max and Victoria Dreyfus Foundation, Inc. The views expressed herein are those of the authors and do not necessarily reflect the views of NOAA or any of its sub-agencies or the Max and Victoria Dreyfus Foundation, Inc.

*Pages; 17, 81, 135, 173, 296:* Double-crested cormorants. Photo: Frozenrope

In memory of my grandmother, Ruth Maude Fisher Gilpin, whose strength and steadfast love has always provided such positive influence; and for my family, for their encouragement and loyal support of career and peregrinations. G.G.B.

For Di and Ken and their genial three; for the crew – Christopher Bruce, Sean Malcolm, and Meghan Quetico Ellington Litteljohn; and for little Eve. B.M.L

For our colleague John Lee, whose invaluable assistance helped to complete this watershed project.
G.G.B. and B.M.L.

On behalf of the Wildlands League, the editors would like to extend their sincere thanks to the following sponsors and friends who generously contributed to the Voices for the Watershed project. These partners come from all regions of the Great Lakes–St Lawrence River watershed. Without their support and dedication to environmental education and conservation, this project would not have been possible.

PROJECT SPONSORS

National Oceanic and Atmospheric Administration's National Undersea Research Center at the University of Connecticut, Avery Point Campus
Max and Victoria Dreyfus Foundation
EJLB Foundation

Brian H. Wheatley, Q.C.
Consumers Gas
McLean Foundation
Upper Canada College

PROJECT FRIENDS

Benjamin Film Laboratories Ltd.
Bluewater Canoes
Federation of Ontario Naturalists
Fuji Photo Film Canada Inc.
Henry's Photographic
Lafarge Construction Materials
H.M. Peacock Foundation
Quebec-Labrador Foundation (QLF/ Atlantic Center for the Environment)

# Contents

Wabakimi Provincial Park
Lake Nipigon
Quetico Provincial Park
Thunder Bay
MINNESOTA
Boundary Waters Canoe Area
Pigeon River
State Forests
Isle Royale National Park
Marathon
ONTARIO
Pukaskwa National Park
Duluth
Apostle Islands National Lakeshore
Lake Superior
Wawa
Keeweenaw Peninsula
Lake Superior Provincial Park
Chequamegon National Forest
Pictured Rocks National Lakeshore
45°
Lady Evelyn Smoothwater Provincial Park
Nicolet National Forest
State and National Forests
Sault Ste. Marie
Lake Temagami
Killarney Provincial Park
Green Bay
Sleeping Bear Dunes National Lakeshore
WISCONSIN
Wolf River
Manitoulin Island
French River
Lake Nippissing
Lake Huron
Georgian Bay
Bruce Peninsula
Lake Winnebago
Lake Michigan
State and National Forests
Muskegon River
Milwaukee
Collingwood
Bay City
Saginaw
Grand Rapids
Goderich
Lake Simcoe
MICHIGAN
Grand River
Toronto
Flint
Chicago diversion canal
Chicago
Indiana Dunes National Lakeshore
Sarnia
Lake St. Clair
Hamilton
Grand River
London
Detroit
Windsor
Welland Canal
Buffalo
South Bend
ILLINOIS
40°N
Pt. Pelee
Lake Erie
Toledo
Illinois River
INDIANA
Maumee River
Cleveland
Cuyahoga River
OHIO
PENNSYLVANI
90°W
85°
80°

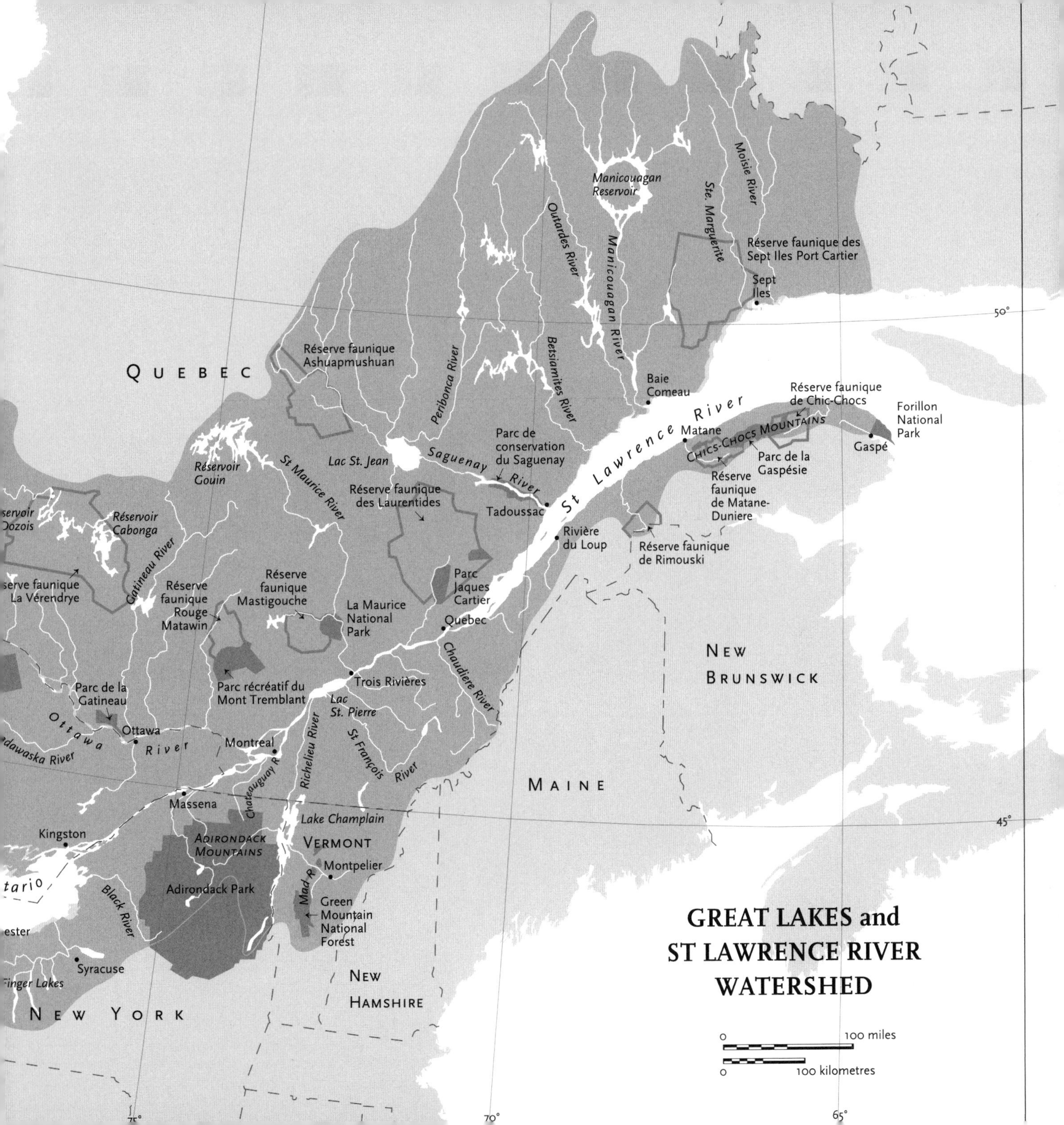

GREAT LAKES and ST LAWRENCE RIVER WATERSHED
QUEBEC
NEW BRUNSWICK
MAINE
VERMONT
NEW HAMSHIRE
NEW YORK
Manicouagan Reservoir
Moisie River
Ste. Marguerite
Outardes River
Manicouagan River
Réserve faunique des Sept Iles Port Cartier
Sept Iles
Réserve faunique Ashuapmushuan
Peribonca River
Betsiamites River
Baie Comeau
St Lawrence River
Réserve faunique de Chic-Chocs
Forillon National Park
Matane
Chics-Chocs Mountains
Gaspé
Parc de la Gaspésie
Réserve faunique de Matane-Duniere
Lac St. Jean
Saguenay River
Parc de conservation du Saguenay
Réservoir Gouin
St Maurice River
Réserve faunique des Laurentides
Tadoussac
Rivière du Loup
Réserve faunique de Rimouski
Réservoir Cabonga
Gatineau River
Réserve faunique La Vérendrye
Réserve faunique Rouge Matawin
Réserve faunique Mastigouche
La Maurice National Park
Parc Jaques Cartier
Quebec
Chaudiere River
Parc récréatif du Mont Tremblant
Trois Rivières
Parc de la Gatineau
Ottawa
Ottawa River
Lac St. Pierre
St François River
Richelieu River
Montreal
Chateauguay R.
Massena
Kingston
Lake Champlain
Adirondack Mountains
Adirondack Park
Montpelier
Mad R.
Green Mountain National Forest
Black River
Syracuse
Finger Lakes
50°
45°
70°
65°
0
100 miles
0
100 kilometres

Northern portage on the height of land. Photo: Frozenrope

# Foreword



On a canoe trip in the rugged country north of Superior, I once portaged over a long ridge indicated by a dotted line on our topographic map as the "Height of Land." The line marked where a drop of rain had to choose whether to flow north through the Arctic watershed to James Bay, or south through wild rivers to the Great Lakes. There I was, poised like that drop of rain, casting my gaze in opposite directions over vast drainages that have accounted for so much history. As it turned out, I headed south, to travel the waters that form the focus of this book.

Under the capable editorship of Gregor Beck and Bruce Litteljohn, experts from the United States, Quebec, and Ontario have contributed their thoughts, concerns, and solutions to the pages of *Voices for the Watershed.* The result is a powerful confluence of voices that brings together the headwater areas I was paddling, the Great Lakes themselves, the mighty St Lawrence, and that great river's outflow in the sea.

Like nature, the book rightly crosses human artifacts such as political boundaries and professional disciplines. It shows us what is at stake and points to a better route ahead if we care enough to choose it. In my view, we have cared so little for so long that now we really have no choice.

*Voices for the Watershed* will do much good for the environmental and human health of this huge, heavily populated, highly industrialized, seriously threatened, but still beautiful part of our world. After all, we cannot expect nature's forgiveness forever.

*Monte Hummel*
*President*
*World Wildlife Fund Canada*

Section 1

# VOICES FOR THE WATERSHED

*"Citizens will continue to be the driving force for the protection of the Great Lakes and St Lawrence watershed. Those living around the region are the ones most directly affected, and who enjoy and value the lakes, rivers, and surrounding lands for their multitude of essential and delightful facets. We have learned that it is only by bringing pressure to bear in a united and simultaneous way on all responsible jurisdictions and polluters that there is hope for protecting the watershed."*

– John Jackson

Photo: Scot Stewart

# Prologue

## *Bridging the Knowledge Gap – Science and Society Working Together for Environmental Understanding*

*Bruce Conn*

I WILL NEVER FORGET THE FIRST TIME I DONNED MY SCUBA GEAR and slipped beneath the water's surface at Tibbett's Point, where the majestic St Lawrence River flows out of the blue depths of Lake Ontario. Down below, the river's floor literally crawled with life – rich, diverse, lovely. The abundance and variety of clams, snails, insects, sponges, sculpins, and other fish was astounding. More than ever before this experience impressed upon me the beauty and complexity of the ecosystem that I had been studying for so many years. And more than ever before, my gut longing was to rush out and do the two things that I enjoy most: to immerse myself in even more research to understand this fascinating environment, and to dedicate more time to conveying my knowledge to anyone who might be enticed to share my love for the river.

To me these two activities are inseparable. How can you possibly conduct research on something without becoming excited about your discoveries, and how can you possibly be excited about discoveries without being driven to share your new-found knowledge with others? This relationship seems natural. Yet in practical terms the natural union between research and education is often broken. The pressures of generating data, producing technical publications, and attracting funding consumes the time of most scientists, leaving them isolated from the public. This isolation is increased by the fact that scientists must be sticklers for accuracy, while journalists and the public often want quick and simple answers to questions that are exceedingly complex and always incompletely understood. Even among academic scientists who regularly interact with students in the classroom, the exhilaration of new discovery often gets lost in the deluge of technical jargon that makes science appear unapproachable to most people.

Yet I am always amazed at how quickly most people become enthralled by even the most mundane of creatures if they are simply taught in simple terms about their beauty and ecological relevance. A good example comes from my own work with caddisflies. These small moth-like insects, known in many localities as shadflies, swarm by the tens of millions along the shores of the St Lawrence River each spring. Their swarms are so dense that they foul porches, smear windshields, produce fish-like odours, and even cause minor allergies in some people. For these reasons, most people who live along the river regard them as pests, despite the fact that the insects never bite, and form nuisance swarms for only a couple of weeks each year. But my research has centred on learning what the caddisflies do during the other fifty weeks of the year. As it turns out, they live as worm-like larvae deep in the river's channel. The larvae are skilled architects, building elaborate houses of twigs and sand grains. They feed on plankton and algae, and are themselves eaten by fish, thus passing nutrients along the food chain that supports a vigorous fishing industry. Caddisflies are among the most important components of the St Lawrence River ecosystem.

But how can I blame people for hating caddisflies when they are exposed only to the nuisance swarms of mating adults and have no way of knowing what the larvae do on their behalf throughout most of the year? One of my responsibilities as a scientist is to help people understand how interesting caddisflies really are, and how critical they are to the environment, and thus to our health and economic well-being. In accepting this responsibility, over the years I have spoken to scores of radio and newspaper reporters, grade-school classes, civic groups, and others. I have never been disappointed with the responses of my audiences. Not everyone goes away loving the spring swarms, but most become more willing to accept the short-term inconvenience in exchange for the long-term rewards.

The time has come for scientists, journalists, and the general public to overcome the barriers to communication and work to restore the natural pairing of scientific discovery and environmental education. Our environment depends on how well we do this. The health of the Great Lakes and St Lawrence River watershed, the world's largest freshwater system, is critical to the well-being of the entire planet. Yet few people appreciate the full magnitude of the natural treasures that we have right in front of us. By doing research, scientists can learn more about these treasures. But today's scientists must go beyond research, even as today's teachers and journalists must strive to develop a better understanding of science. Together, we must bridge the ivory tower, the classroom, and the public square. Only by cooperating closely with teachers and journalists can scientists ensure that what they learn will be conveyed to the public. And it is the public, working through economic and political action, who ultimately must be the advocates and caretakers of our natural world.

St Lawrence River near Tadoussac, Quebec. Photo: Gregor G. Beck

# Introduction

*Bruce Litteljohn and Gregor Gilpin Beck*

IN MAY 1535 FRENCH MASTER MARINER JACQUES CARTIER set sail from St Malo in Brittany. By command of his king, he was to discover the western lands beyond Terres Neufres, which we now know as Newfoundland.

Cartier's voyage took him into uncharted and dangerous waters. By mid August, however, he had navigated his way deep into the estuary of the St Lawrence River. Proceeding upstream, he and his crew paused to admire beluga whales near Tadoussac. In September they arrived at the magnificent site now occupied by Quebec City. Continuing to battle strong currents, Cartier took less than a month to bring his boats to the present location of Montreal. Here he named and climbed Mount Royal, learned of vast seas further inland, and – blocked by the turbulent Lachine Rapids – headed back to Quebec where he and his men wintered before returning to France.

Cartier's accomplishment was of enormous importance. To European newcomers he handed the key to the interior: the major water entry to the North American heartland. This, of course, was the St Lawrence–Great Lakes system with its thousands of tributary rivers and lakes fanning out to drain a vast watershed.

Other French explorers followed in Cartier's wake. Samuel de Champlain founded a permanent settlement at Quebec in 1608. In the following year, led by Huron Indians, he ascended the Richelieu River to traverse a huge lake which was given his name – a major southern extension of the Great Lakes–St Lawrence axis. By the time of his death in 1635, he and his followers had explored the Ottawa River, Lake Nipissing, Lake Simcoe, and part of Georgian Bay on Lake Huron. They had also crossed Lake Ontario to what are now known as the Finger Lakes in upstate New York and travelled west to Green Bay on Lake Michigan and possibly as far as the eastern end of Lake Superior. Much had been done in very few years to reveal the core of the watershed.

By 1700 the Great Lakes and numerous tributary rivers and lakes, such as Lake Nipigon in the far northwest, were well known to the French. Many explorers had crossed the height of land into adjoining watersheds such as those of the Mississippi River and Hudson Bay. In doing so they had done much to define the boundaries of a freshwater system that is staggering in its magnitude. Both in the size of its core lakes and the volume of water that flows and eddies through the drainage basin, nothing on earth can match it. In a world beset with growing deserts and water shortages that threaten the well-being of many millions of people, we are incomparably rich in water. Our challenge is not to find water but to keep it clean and usable for ourselves and all other forms of life that depend on it.

How much do we have? Lake Superior alone contains close to 10 percent of all the free-flowing fresh water on the earth's surface. Taken together, the Great Lakes hold about 20 percent of the world's available fresh water. When we add the St Lawrence and all the feeder lakes and rivers, some of which (Lakes Champlain, Nipigon, Nipissing, and Lac Saint-Jean) are very large, we are looking at approximately 25 percent of the earth's fresh surface water. While this is indeed remarkable, we must not forget that only about 1 percent of this amount is actually renewable through precipitation; the balance is a remnant of glacial meltwater.

Rivers and lakes are the natural focus of the huge region introduced in this book. They are the unifying feature as well as the most obvious geographical expression of the Great Lakes–St Lawrence watershed.

In terms of human history they have played a vital role. They have provided the main highways and byways of the original native peoples and of the explorers, fur traders, and settlers who arrived from Europe. They provided clean water for drinking and fresh, untainted fish in abundance for food. For the new European-based economic system they made possible the transport of trade goods and furs and, later, logs from the vast forests that clothed the land between rivers and lakes. Most of our cities and towns grew up along the shores of the main axis of the watershed: Tadoussac, Quebec, Trois Rivières, Sorel, Montreal, Kingston, Rochester, Toronto, Hamilton, Niagara Falls, Buffalo, Erie, Cleveland, Toledo, Detroit, Windsor, Sarnia, Gary, Chicago, Milwaukee, Green Bay, Sault Ste Marie, Superior, Duluth, and Thunder Bay, to mention just a few. And soon the rapids and falls were by-passed by ship canals to make possible the movement of wheat and iron ore from the interior and incoming goods from overseas. The heartland of North America, to which Cartier led the way, was now open to water-borne world trade.

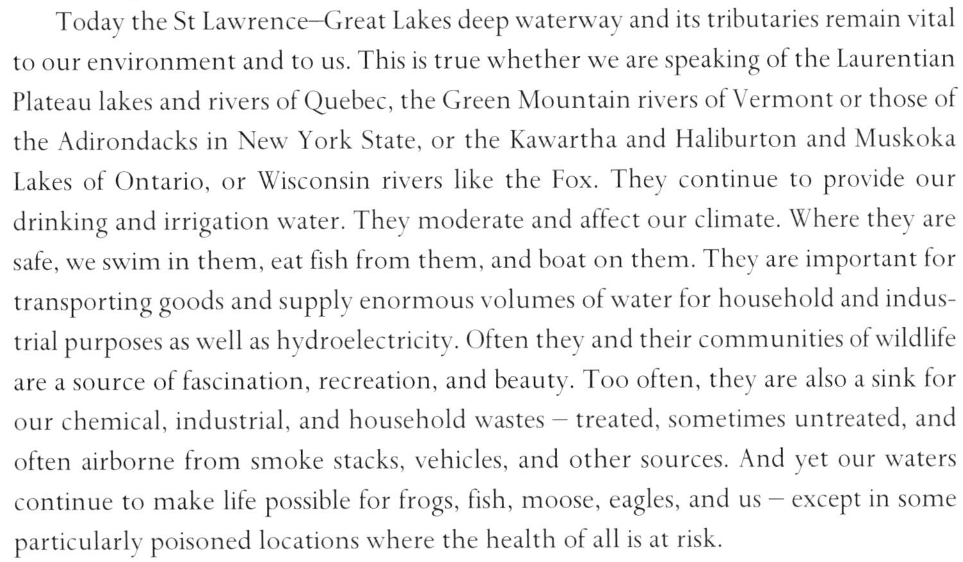

Today the St Lawrence–Great Lakes deep waterway and its tributaries remain vital to our environment and to us. This is true whether we are speaking of the Laurentian Plateau lakes and rivers of Quebec, the Green Mountain rivers of Vermont or those of the Adirondacks in New York State, or the Kawartha and Haliburton and Muskoka Lakes of Ontario, or Wisconsin rivers like the Fox. They continue to provide our drinking and irrigation water. They moderate and affect our climate. Where they are safe, we swim in them, eat fish from them, and boat on them. They are important for transporting goods and supply enormous volumes of water for household and industrial purposes as well as hydroelectricity. Often they and their communities of wildlife are a source of fascination, recreation, and beauty. Too often, they are also a sink for our chemical, industrial, and household wastes – treated, sometimes untreated, and often airborne from smoke stacks, vehicles, and other sources. And yet our waters continue to make life possible for frogs, fish, moose, eagles, and us – except in some particularly poisoned locations where the health of all is at risk.

The rivers and lakes are like the essential arteries and veins of a body. That body is the entire watershed which includes all the land from which water drains to find its way into the Great Lakes and St Lawrence River. A glance at the watershed map indicates that this is a huge area which includes parts of nine states (Minnesota, Wisconsin, Michigan, Illinois, Indiana, Ohio, Pennsylvania, New York, and Vermont) and major

portions of the provinces of Ontario and Quebec. In total, the drainage basin, including both land and water, is about 1,344,000 km$^2$ (518,500 square miles) in extent.

While this area – more than twice the size of France – is tied together by its drainage pattern to form a geographical unit bounded by the encircling height of land, it is also an area of enormous natural and ecological variety. To the north of the St Lawrence and Great Lakes, the dominant geological feature is the rugged Precambrian or Canadian Shield. Characterized by thin soils and ridges of hard, exposed rock, it is also distinguished by an intricate pattern of hundreds of thousands of lakes, swamps, streams, and rivers. Within the watershed, the shield's vegetation ranges from mixed forests of maple, oak, birch, and white and red pines in the south to the boreal forest in the more northerly portions where spruce, balsam fir, and poplars are the dominant tree species. This boreal component includes the major portion of the Quebec segment of the watershed north of the St Lawrence. South of the great river, the Precambrian Shield also extends into New York State to form the Adirondack Mountains.

The Great Lakes–St Lawrence Lowland is in strong contrast to the rocky shield. This area of deep, fertile soils offers a much gentler landscape of plains and rolling hills. Located immediately north of Lakes Erie and Ontario, it also includes the Ottawa Valley and narrow plains along both shores of the St Lawrence River, mainly south-west of Quebec City. Once blanketed by forests of massive pines, oaks, maples, and beech, almost all of this area has been cleared for agriculture and urban development. At the same time the vast majority of its biologically rich wetlands have been drained or filled for the same purposes. Biodiversity has been radically reduced.

While there are lowlands along some of the southern shore of the St Lawrence, much of the area is mountainous. The Green Mountains of Vermont, for example, fall partially within the watershed and are part of the Appalachian chain that extends north-east up the Gaspé Peninsula and includes the Chic-Chocs Mountains and the Notre Dame Range named by Cartier in 1535. Much of the forest cover here is boreal or Acadian, with conifers predominating.

West of the Adirondacks and the Finger Lakes of upper New York State, the watershed – partially bounded by the Allegheney Plateau – becomes a far less rugged area of slow-flowing rivers and plains that are largely given over to agriculture. From the eastern end of Lake Ontario right around to Green Bay on Lake Michigan this is mainly the case, with rugged northern Michigan being an exception. This predominantly lowland area is also the location of major urban concentrations such as those with Cleveland, Buffalo, Detroit, and Chicago at their cores. Most of the forests and wetlands encountered by Champlain and other early explorers are gone. The Iroquoian and Algonkian-speaking natives who preceded them would find little to recognize in this much manipulated landscape. The Carolinian forests of the southern portion (which included black walnut, butternut, locust, hickories, ash, and oak) are also largely gone, as is most of the original mixed forest of central Michigan and Wisconsin. Only in the northern upland areas of Michigan, Wisconsin, and Minnesota do extensive forests and wetlands remain in the United States portion of the watershed. However, even in these wooded areas little old-growth forest remains.

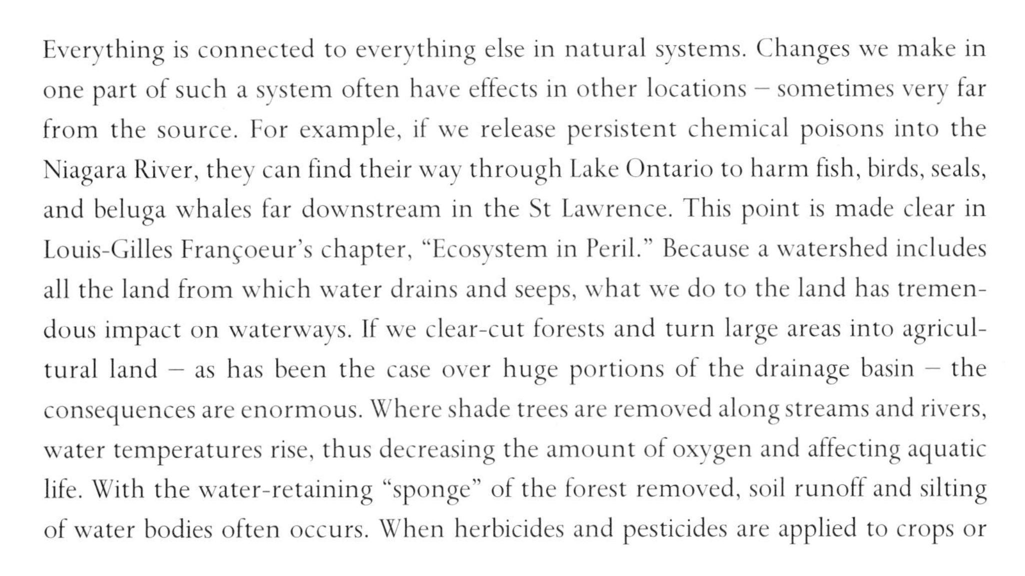

Everything is connected to everything else in natural systems. Changes we make in one part of such a system often have effects in other locations – sometimes very far from the source. For example, if we release persistent chemical poisons into the Niagara River, they can find their way through Lake Ontario to harm fish, birds, seals, and beluga whales far downstream in the St Lawrence. This point is made clear in Louis-Gilles Françoeur's chapter, "Ecosystem in Peril." Because a watershed includes all the land from which water drains and seeps, what we do to the land has tremendous impact on waterways. If we clear-cut forests and turn large areas into agricultural land – as has been the case over huge portions of the drainage basin – the consequences are enormous. Where shade trees are removed along streams and rivers, water temperatures rise, thus decreasing the amount of oxygen and affecting aquatic life. With the water-retaining "sponge" of the forest removed, soil runoff and silting of water bodies often occurs. When herbicides and pesticides are applied to crops or forests, they find their way into the water, as do urine and faecal matter from farm animals. With the forests gone, stream and river flooding increases. And there are other negative consequences. In fact, the quality of our water provides a useful index to our land-based activities, to the ecological awareness of our society, and to the environmental sensitivity of urban, industrial, and agricultural development.

Our destructive capacity where our lakes and riverine systems are concerned has increased, in spite of efforts to restore water quality in some places. In the days of Cartier and Champlain, French explorers travelled upstream towards the height of land through vast forests teeming with wildlife. They could dip their cups over the gunwales of their canoes and drink from the rivers and lakes without a second thought. This is not the case today – especially in the heavily populated areas which encircle the Great Lakes and line the St Lawrence River. Unlike the early explorers, we often look upstream out of concern for what may be flowing toward us in the way of pollutants and toxins. And, rather than discovering new water routes, we must be concerned with saving and restoring the natural integrity of our rivers and lakes and the watershed lands they drain. But how to go about this as population and development grow and environmental agencies suffer major cuts in spending and personnel?

First, it seems to us, we have to be aware of our "ecological address" – the location of our home place within the *natural* environment. As two of our contributors, Peter Lavigne and Stephen Gates, write, "We all know our mailing address, but how many people know their ecological address or their watershed address?" They go on to indicate that "knowing your ecological address, your watershed address, means knowing your relationship with the waterways around you. Where does your drinking and washing water come from? What affects its quality and quantity? Where does it go when you are done using it or when you flush your toilet? What local sub-watershed do you live in, and to what major watershed do you belong?"

These are important questions. While it is useful to know where the local grocery store is located, it is vital to our health and survival (and the survival of other living things) to know where we fit in, and how we depend upon and how we alter our natural environment. This in turn requires some knowledge of how ecosystems work, a topic dealt with in Anne Bell's chapter, "Frog Reflections: The Ecosystem Approach to Conservation." This kind of knowledge is sometimes strengthened by aesthetic, emotional – even spiritual – bonding with areas. Painter Jeff Miller's chapter on the Algonquin dome reflects this multi-faceted approach. From such an ecological, aesthetic, and emotional response to place can come a deep-seated environmental ethic against which human intervention in nature is measured – both at our home location and elsewhere.



For more than 45 million of us, our ecological address is the Great Lakes–St Lawrence River watershed. However, our more immediate interest and concern is likely to be focused on the sub-watershed in which we make our home – whether it be that of the Kalamazoo River in Michigan or the Kaministiquia in Ontario. These and hundreds more are all interrelated and important parts of the greater watershed, to which this book offers an introduction.

How do we learn about our local rivers and lakes that we so often simply take for granted? One way is to go and walk their shores and headwaters, and turn to them for recreation. This can do much to help us know our ecological address and its significance. But it won't tell the whole story, for there is much about water quality that is invisible. For greater knowledge we look to scientists, environmental journalists, and educators.

This book is not intended for environmental experts or for scientists. There are specialized publications for those professionals. For this reason, journalists who are trained to write for the general public are important contributors to this work. We are fortunate that Michael Keating (formerly of the *Globe and Mail*), Rae Tyson (*U.S.A. Today*), and Louis-Gilles Françoeur (*Le Devoir*, Montreal) have written major essays for *Voices for the Watershed*. They are joined by three other environmental journalists, Phillip Norton, Brad Cundiff, and Alec Ross. Environmental writers as well as their colleagues in television and other media help spread important information and ideas to potentially huge audiences that represent all ages and sectors of the population.

Educators too are taught to get information to students in clear and interesting ways. Whether at preschool or high school, college or university, teachers play a vital role in fostering ecological understanding and encouraging active participation in all sorts of environmental activities. It is tremendously heartening to see young children planting trees and teenagers cleaning up shorelines, and to know that college and university students are monitoring the water quality of streams. Important life lessons are learned through these hands-on projects and in the classroom, and they are shared with parents, siblings, and friends. Among the educators who have written here are J. Douglas Blakey, principal of Upper Canada College in Toronto; Scot Stewart, a science teacher from Marquette, Michigan; and Kevin Coyle, president of the National Environmental Education and Training Foundation in Washington, D.C.

Unlike journalists, scientists (upon whom we depend for leading-edge knowledge of what is happening to the watershed environment) often write in a highly technical manner which is difficult for laypersons to understand. Bruce Conn both recognizes and surmounts this obstacle in the prologue, "Bridging the Knowledge Gap." The prologue also underscores one of the chief goals of this book, which is to make the discoveries of front-line research scientists understandable to those who are not familiar with the specialized language of science. By so doing, major environmental problems can be understood by the general public, as can possible solutions to such problems. Hallett J. Harris and Val Klump write, for example, of the problems associated with agricultural non-point source pollution; Michel Letendre discusses the threats to biodiversity in the Montreal region; Peter Ewins summarizes the problem of bioconcentration of toxins in fish-eating birds in the Great Lakes; and Jean Rodrigue, Louise Champoux, and Jean-Luc DesGranges explain how biological indicator species in the St Lawrence are being used to track pollutants.

Once we know our watershed address and have informed ourselves about environmental problems and solutions, we are in a position to act. Without action there will be no improvements or solutions.

Where action is concerned, we cannot simply expect governments and government agencies to do the job. For example, Nadia Ménard writes in these pages of the Saguenay–St Lawrence Marine Park in Quebec. However, there would be no such park had it not been for public concern with the fate of the now-endangered beluga whales in the park area and the hard work of non-governmental conservation groups such as World Wildlife Fund and Quebec conservation groups.

This is not to suggest that government researchers and administrators, both in the United States and Canada, have not made very significant contributions to our knowledge. They have; but, unfortunately, knowledge of problems has not always led to effective action. There are many governments and government agencies in the watershed. Sometimes their jurisdictions overlap, and sometimes they do not march to the same drummer. In recent years, too, there have been severe reductions in government staff and funding devoted to scientific research and action required to investigate, monitor, and improve water quality and associated environmental problems. It is fortunate that some of the government people who lost their jobs, as Gail Krantzberg points out in "Making the Lakes Great," have chosen to work on their own time and to cooperate with grassroots citizen groups about which John Jackson writes. For Jackson, who is former president of Great Lakes United, an international non-governmental organization, "Citizens will continue to be the driving force for the protection of the Great Lakes and St Lawrence watershed."

Jackson's opinion is based on long experience. It is reinforced in the Châteauguay River essay by Serge Bourdon and Phillip Norton, in Elliot Gimble's writing about friends of Vermont's Mad River, and in many other places in this book. Scientific expertise and government agencies can be of great importance, but most of all we need informed and concerned citizens who are prepared to take action to protect or restore the natural quality of their neighbouring rivers and lakes and the lands surrounding them.

One of the best ways to translate your concern into really effective action is to join and otherwise support conservation and environmental groups. These may be very

small organizations mostly dealing with the hands-on cleaning up of local streams or lakeshores. Some are large and international in scope, such as the Nature Conservancy, Great Lakes United, or World Wildlife Fund. Others include l'Union québecoise pour la conservation de la nature, the Sierra Club, Pollution Probe, the Canadian Nature Federation, the Cenozoic Society, le Fonds pour la sauvegarde des espèces menacées, the Federation of Ontario Naturalists, the Wildlands League, the National Audubon Society, the Quebec-Labrador Foundation, Friends of the Earth, American Rivers, and the Canadian Parks and Wilderness Society's Ottawa Valley and Wildlands League chapters, and so on. Many of the larger organizations have local chapters. *All are important.* Just as one example, World Wildlife Fund Canada, in conjunction with the Federation of Ontario Naturalists, the Wildlands League, l'Union québecoise pour la conservation de la nature, and many regional groups, is currently completing an extremely well-researched and broadscale "Endangered Spaces" campaign. Already quite successful in Ontario, the campaign will ensure the protection of a wide range of natural areas on the Canadian side of the watershed when Quebec sets aside lands. This kind of work absolutely needs your support – as do countless smaller-scale, but still vital, environmental and conservation projects. Without them, there would be little government action, and the watershed – the core of which is already dangerously polluted and abused – would be in far worse shape.

The work of striving for a healthy environment in this heavily populated area will, of course, never end. There will be new sources of pollution, grandiose and ecologically devastating mega-plans to divert the Great Lakes and other waters, poorly controlled developments that threaten precious habitat, massive water-withdrawal proposals, and the like. We need the protected places about which Jerry De Marco writes, and we need the Remedial Action Plan committees with which Louise Knox in Hamilton deals. And they all need us.

Our purpose in making this book has been to emphasize natural links and common challenges and to draw attention to the fact that the drainage basin is held together by its web of lakes and rivers. For this reason, we have taken the highly unusual approach of introducing the *entire* Great Lakes and St Lawrence River watershed. There have been many books and articles about either the St Lawrence or the Great Lakes, as if they were separate from one another. They are not. As the early explorers came to realize, this is a single water-welded, interrelated, and interdependent area of enormous significance. For too long people have been trying to understand and address environmental issues based on political boundaries and local bureaucratic policies, and not on the all-important natural ecological systems.

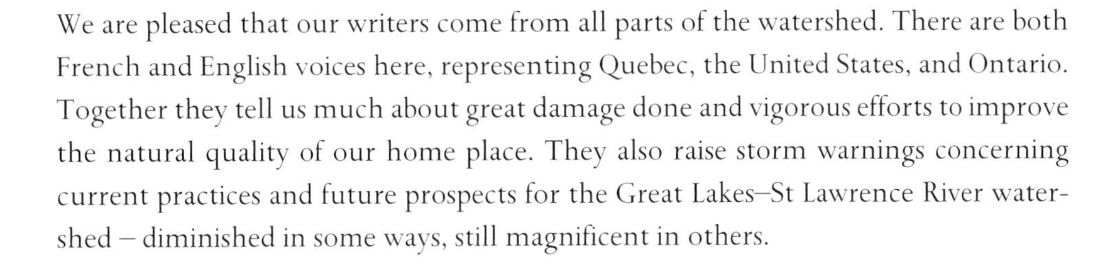

We are pleased that our writers come from all parts of the watershed. There are both French and English voices here, representing Quebec, the United States, and Ontario. Together they tell us much about great damage done and vigorous efforts to improve the natural quality of our home place. They also raise storm warnings concerning current practices and future prospects for the Great Lakes–St Lawrence River watershed – diminished in some ways, still magnificent in others.

Moose. Photo: Frozenrope

# Headwaters and Upland Regions

Genesee River, New York State. Photo: Frozenrope

# Highland Headwaters

## *Getting a Good Start along the Height of Land*

*Bruce Litteljohn*

*"Natural integrity and headwaters protection have too often been ignored on our journey to the sea. Downstream, the problems are compounded, but the highland headwaters are where it all begins. If the vital rivers that spill into the Great Lakes and St Lawrence are to thrive and deliver health and life ... they need a good start in life."*

THOUGHTS OF HEADWATERS call up the wonderful imagery of Bill Mason's film adaptation of *Paddle to the Sea*. The tale begins in the hard-rock headwater hills north of Lakes Superior and Nipigon. Here a lovingly carved model canoe, complete with a native sternsman named Paddle, begins its long voyage through the Great Lakes and the St Lawrence to the Atlantic.

Produced by the National Film Board of Canada and based on the classic children's story by American writer Holling C. Holling, this brilliantly filmed romance is a journey through the world's greatest freshwater system. It has engaged the minds and hearts of children, as well as those adults whose imagination, sense of adventure, and love of natural beauty and grandeur have survived their own journeys. Who can forget the thrill of anticipation as Paddle, in the melting snow of a northern springtime, slides tentatively toward a highland stream and then into its tumbling, jubilant waters. From that point, Paddle is committed, and so are we.

From the pristine headwaters, Paddle rides the currents south through Superior and then to the heavily populated lower Lakes. He encounters the industrial heartland of North

America with its huge, steel-sided ships, fouled waters, and angular cityscapes. It is a voyage from wilderness to an overcrowded, utilitarian, built-and-exploited environment where forests and wetlands have been destroyed by agriculture and urban sprawl. To find our way back to unspoiled headwaters conditions – to metaphorically reverse Paddle's journey and regain the high ground – poses a tremendous challenge.

Well over a half century has passed since Holling wrote his book in 1941. It predated the postwar introduction of synthetic pesticides and the massive use of chemical fertilizers. Plastics and the vast expansion of the petrochemical industry, with their associated environmental degradation, had yet to make their impact, and nuclear power was still an infant wartime project. And only a small fraction of the more than 45 million people who now inhabit the Great Lakes–St Lawrence watershed[1] lived there when Holling wrote Paddle to the Sea. The Second World War triggered enormous economic productivity and industrial growth. Only recently have we realized the magnitude of the environmental damage done by sweeping technological and industrial upsurge spurred on by the profit motive, seductive advertising, and consumer appetites. Only recently have we noted the loss of free-running brooks from which we can safely drink, towering forests with "shadowy sun-pierced silences,"[2] and all the life, health, and beauty that they represent. Sun-pierced silences have become harder to find even in many of the headwaters areas defining the boundaries of this immense watershed. Circling the outer limits of the basin from the Chic-Chocs and Notre Dame Mountains of Quebec's Gaspesie and Bas Saint-Laurent regions, southwest to the Green Mountains of Vermont and the Adirondacks of northern New York State, we commence a long, sometimes lovely, but often sobering, journey.

We begin in Quebec, south of Fleuve Saint-Laurent, where action to protect true wildlands is far from complete. Here, east of Rivière-du-Loup, the northern Appalachians brush the St Lawrence, and large provincial wildlife reserves are found in vital headwaters locations. But these blocks of green on the map are misleading. The Reserve faunique de Matane-Dunière and that of Rimouski, for example, are extensively logged, and the sun-pierced silences are frequently broken by the roar of chain saws, skidders, and hauling trucks. These are "multiple-use" areas rather than nature sanctuaries. They attempt, with limited success, to combine recreation with a variety of commercial and industrial uses. This situation is partially overcome by Parc de la Gaspésie, which protects the height of land at the extreme eastern end of the watershed and Parc de Frontenac south of Quebec City. However, only about 1 percent of the total area south of the St Lawrence is protected as roadless preserves where hunting, mining, and extractive logging are forbidden. To help correct this pathetic situation, the World Wildlife Fund and l'Union québecoise pour la conservation de la nature (supported by several other groups) have recently published a very useful map indicating candidate natural areas for legal protection. Many of these are headwaters areas.

Yes, quiet, clean places can be found in protected parts of Vermont's Green Mountains, and by streams that tumble down to Lake Champlain and from there to the St Lawrence. Vermont is a largely rural state, seemingly highly conscious of conservation. In fact, it provides a good example of the re-greening of the American East and its recovery from years of logging and destructive agriculture (c. 1830-1860). With the opening of the Erie Canal in 1835, Vermont's economy began to shift to service industries and manufacturing, but the restoration and return of wildlife has been a long process. These are ecologically young, replacement forests and not virgin woods,[3] and natural regeneration remains at risk from the current wave of developers and loggers. The Green Mountains (including the large National Forest) are pierced by roads, tourist development continues to grow, and commercial logging is fairly extensive. But although little true wilderness or old-growth forest remain, many headwaters are protected. This is the case even though highways run along several major rivers such as the Missisquoi, Lamoille, and Winooski.

Lake Champlain, shared by New York State, Vermont, and Quebec, is one of the largest of the height of land lakes. Only twenty-nine metres (ninety-five feet) above

sea level, it flows into the St Lawrence at Sorel via the Richelieu River. The lake suffers from many of the environmental problems of the Great Lakes: exotic species like Eurasian milfoil and zebra mussels; severe phosphorus pollution in several areas; loss of wetlands; waterborne pathogens such as *Giardia*; and toxic pollutants. Growing pressure from visitors and over 600,000 residents in the drainage basin is also a factor. Yet it is these problems that have led to the eyebrow-raising claim that Lake Champlain should be numbered among the Great Lakes. In truth, the debate was about accessing federal research funds to clean up the lake.[4] In 1998 Senator Patrick Leahy succeeded in having Lake Champlain promoted to Great Lakes status for the purpose of funding ecological research under the National Sea Grant Program. Champlain's water quality has improved over the past twenty years, and remedial work continues through the Lake Champlain Basin Program and partners such as the Nature Conservancy.



West from Lake Champlain, the southern extremity of the vast Precambrian Shield crosses the St Lawrence at the Thousand Islands and forms the mountainous Adirondack State Park. From this beautiful upland, brooks swell to rivers that tumble down to Lake Ontario, the St Lawrence, and Lake Champlain. In those areas of the park that are truly protected (only about 20 percent has been classified as wilderness, despite the 1892 statute that set aside all the public lands to be "forever wild"), hundreds of headwater streams and lakes that look pristine are assaulted and made nearly sterile by the insidious drift of acid rain and snow. As park naturalist Mike Story indicates in *A Natural History of the Adirondack Park, N.Y.*, "By 1980 approximately 60% of all the lakes above 2000 feet in elevation were at critical levels of acidity." Ten years later, 270 of the park's 2,800 lakes were acid-dead, and the problem continues to grow. Added to this disaster is mushrooming population pressure. The road network has grown apace, with developers standing in the wings. Indeed, from 1972-92, 22,000 homes were built within the park. Numbers of year-round residents grew from about 6,200 in 1892 to roughly 135,000 in 1992 – plus 210,000 seasonal residents and 10 million annual visitors. Loggers, too, play a large role, both by cutting substantial amounts of timber and, recently, selling portions of their holdings to those who would build on them.

Fortunately, others are standing ready in the wings as well. The "Algonquin to Adirondacks (A2A) Conservation Initiative," a partnership of individuals and groups including the Canadian Parks and Wilderness Society and The Wildlands Project, proposes a large cross-boundary landscape where natural connectivity is maintained through cooperative private land stewardship. This interesting concept, based on a careful study of conservation biology, would see the Adirondack dome connected to Algonquin Provincial Park in Ontario by means of a natural corridor through which animals could move. The "Greater Laurentian Wildlands Project" is developing a long-term conservation strategy for an even larger area.[5]



The journey south from the hard granite and gneiss of the Adirondack Mountains takes us downhill to the lake plain area north and east of Oneida Lake. Near Rome, N.Y., the divide skirts the headwaters of the Mohawk-Hudson River system and crosses the Erie Canal (one of several diversions). For a time we are in a landscape of farms, swamps, and relatively high population. Here there is urban and industrial pollution and runoff from pasturelands and low-intensity farming. A pamphlet by the New York Office of Parks for the Old Erie Canal Park invites the public to "canoe the legendary waters of the Old Erie Canal," but adds, "CAUTION. Swimming and wading in the canal are discouraged. If water contact is made, proper hand washing is strongly recommended before eating or smoking." I'm glad Paddle to the Sea didn't take this short-cut to the Atlantic.

As the land rises south of Syracuse, we encounter outriders of the great Appalachian Mountain chain extending from Georgia up to the Gaspé Peninsula and to the "Rock" of Newfoundland. Elevations rise rapidly. The height of Oneida Lake (part of the Erie Canal system) is 370 feet (112 metres) above sea level. To the southwest Otisco Lake is 1191 feet (363 metres) above sea level. Adjacent highlands such as Bear Mountain tower 450 metres higher. This feels more like real headwaters country, and here we find the creeks and streams that supply the beautiful Finger Lakes. These

in turn drain into Lake Ontario through the Seneca and Oswego rivers. This is voluptuous hill and valley country, the long glacier-gouged lakes interspersed with massive drumlin hills. However, the slopes are not too steep and rocky to deter farming in this agricultural cornucopia.

Farming leads to runoff of chemicals, urine, and feces; other problems include housing development, swamp destruction, recreational pressures, and insufficiently rigorous zoning standards. As elsewhere in the Great Lakes–St Lawrence River watershed, political jurisdictions ignore ecological boundaries. A fierce tradition of home rule among the communities and counties of New York State complicates mapping a common strategy to conform to natural units and protect the environment. However, things can change. Betsy Landry, program coordinator of the Water Resources Board of the Finger Lakes–Lake Ontario Watershed Protection Alliance (FLOWPA),[6] says management plans for most of the lakes are in place, and inter-municipal cooperation is growing.

The Finger Lakes also suffer from exotic intruders: the now-omnipresent zebra mussel, water milfoil, spiny water flea, and the dangerous water chestnut, as well as seeping septic tanks from the many cottage and home owners encircling the lakes. But as Betsy Landry puts it, "We'll get there one step" – or one lake, or one community – "at a time."

West from the Finger Lakes the divide follows the edge of the Appalachian and then the Allegheny uplands. Here there is little cause for optimism. Increasingly, this is a built, industrialized, and intensively farmed environment with major pollution problems. Closer to Lake Ontario, a striking anomaly extends the watershed far to the south: the Genesee River, which rises in Pennsylvania and flows out to Lake Ontario at Rochester. While it suffers from agricultural pollution in its upper reaches and industrial insults towards its outflow, it has cut a spectacular shale and sandstone gorge in the vicinity of Letchworth State Park. Here we find the "Grand Canyon of the East," almost six-hundred feet deep and frequented by kayakers and tourists. Here we also find the huge Mt Morris dam, the largest east of the Mississippi, constructed by the U.S. Army Corps. of Engineers. The need for this colossus is difficult to fathom. Built for flood control, it held back absolutely no water during autumn of 1999.

The Mt Morris dam raises the key issue of impoundments along other headwaters rivers. Startling in their number, environmental impact, and – sometimes – size, these dams are for flood control, hydropower generation, irrigation, and facilitation of navigation or recreation. Others provide backyard and livestock ponds and artificial fishing holes. Some are useful and, given our thoughtless history of building on flood plains and denuding riverside forests, necessary. Others have become worse than useless through silt deposits behind them, artificial and destructive water levels, and lack of funding for major, increasingly necessary maintenance. Many of these structures, great and small, are in private hands. A large proportion block sensitive headwater streams.



Tim Tiner, writing of Ontario's almost two thousand dams that exceed two metres in height, notes that "even small impoundments are incredible shocks to riverine ecosystems, disrupting the geophysical processes and plant and animal communities that evolved with the flow and natural cycles of each stream."[7] This means that oxygen levels, sediment patterns, and water temperatures are altered and that species disappear with the loss of natural habitats. Fish and invertebrate migrations are blocked, leading to death or "genetically isolated populations."[8] Vascular surgeons are familiar with the concept: block the flow of oxygen-rich, life-giving blood, and bad things happen.

The case for untrammelled, free-flowing rivers has been brilliantly made in David M. Bolling's *How to Save a River*, which looks to solutions as well as indicating and explaining riverine problems. There may be well over 2.5 million dams in the United States alone, Bolling notes, and "less than 1 percent of the nation's river miles are protected in their natural state." While some dams have been beneficial, he argues that far too many have been built for no justifiable reason. "Between 1962 and 1968, more than 200 major dams were completed in North America each year and, during the same

period, smaller dams were being constructed at a rate of more than 2000 per year." Bolling's eloquence is as compelling as his knowledge: "People who love rivers don't usually need technical explanations or elaborate inventories to explain or defend their value. A river is worth saving for what it manifestly is: a corridor of water, rock and land, a zone of life, a place of inexpressible beauty constantly reshaping itself. But the value of rivers exceeds anything most of us can imagine – it encompasses the very essence of planetary life. Healthy rivers are so important they define, in many respects, the health of the planet." [9]



Bolling's concern for the environment leads him to favour natural rather than structural flood controls. This means leaving rivers with their meanders and wetlands intact, surrounding forests unharvested, and flood plains undeveloped. Given such an intelligent approach, our rivers would have an enormous capacity to absorb rainfall and check runoff. This would also be much cheaper than building dams. Referring to the tragic Mississippi River flood of 1993, Bolling notes, "Dams and levees don't stop floods, they just make floods less frequent and more devastating." His observations are borne out by recent floods on the Chateauguay and Saguenay Rivers. With such thoughts in mind, conservation authorities have begun to think about dam removal or partial removal or reconstruction, including channel bypasses for aquatic life. The pendulum is beginning to swing toward rehabilitation and the "freeing" of some rivers.

From the Genesee River in New York State, the divide extends west and south not far from the Portage Escarpment of the Allegheny (Appalachian) Plateau. The escarpment rises sharply from the lowland, and east of Cleveland it is less than eight kilometres (five miles) from Lake Erie. The height of land is a bit further inland among the rolling hills of the Appalachian uplands. This is an area of fertile soils, intensive agriculture, and urban concentrations such as Akron, an increasingly built and industrialized environment. The lakefront cities of western New York, Pennsylvania, and Ohio have been working at rehabilitation of their waterfronts, and there have been improvements such as the Cuyahoga Valley National Recreation Area – the Cuyahoga River at Cleveland is not likely to catch on fire again as it did in 1969. But population growth and agricultural runoff challenge conservationists, and there are no adequately large protected areas guarding the headwaters. From Cleveland through Toledo, Detroit, Chicago, and Milwaukee we enter a vast lake plain. Prior to 1850 about 4,000 square kilometres (1,500 square miles) of this area were covered by the Great Black Swamp, but by 1900 clearing, draining, filling, and conversion to farmland had reduced it to a mere 100 square kilometres (forty square miles)! It has continued to shrink, partly because of increasing development pressure. However, some private hunt clubs and government preserves have saved vestiges of the great wetland.



South of the flatlands the divide that tilts the rivers toward the Great Lakes is barely discernable. The Sandusky, Maumee, Portage, and St Joseph rivers flow slowly through a region of corn belts, pasture land, intensive general farming, and many sizeable towns. Transportation corridors and pipelines abound, and there is not a single protected area of any size along the roughly 640 kilometres (four hundred miles) of height of land between Cleveland and Chicago. In the absence of commitment on the part of government agencies, informed private stewardship of the land is vital here.

The divide again approaches close to water's edge near Chicago, the largest Great Lakes city. At one point it is within eight kilometres (five miles) of the shore. At another it is only about six feet above lake level. Here the height of land has been breached. In 1820 the Chicago River flowed from it into Lake Michigan. This was changed in 1900 because the river was carrying typhoid and diphtheria to within easy reach of Chicago's drinking water intake. Rather than cleaning up the river (described as "defiled and putrescent with sewage and filth"[10]), the city decided to reverse its flow. The Chicago Sanitary and Ship Canal now flows out of Lake

Michigan, into the Mississippi River drainage basin. Revolting raw sewage solids no longer wash up on toney Gold Coast beaches; the stuff (better treated now) goes south. However, other area rivers such as the Calumet continue to flow north through Chicago into Lake Michigan, so the city remains a somewhat compromised part of the great watershed.

To be fair, much has been done to clean up the waterfront of the megalopolis and make it attractive. But, this tends to be limited to select areas, and the greater urban areas such as East Chicago and Gary, Indiana, remain smoky, polluted, and ugly. The disparity between rich and poor is particularly evident along lakeshores. Within Cleveland or Chicago or Detroit one sees downtrodden and filthy places; just beyond, miles and miles of baronial estates line the shores.



North from Chicago to Milwaukee, the height of land remains within about thirty kilometres (twenty miles) of the Lake Michigan shore. Most of this region is heavily urbanized, primarily urban runoff territory. Here the Waukegan Harbor and Milwaukee Estuary were identified as "areas of concern" in need of strong remedial action. This was dramatically revealed by the 1993 outbreak of waterborne disease that sickened about 400,000 people in Milwaukee, hospitalized more than four thousand, and killed 103. The apparent cause was the tiny *Cryptosporidium* pathogen frequently found in livestock feces. With farmland streams and rivers almost invariably cleared to their edges, excrement, pesticides, herbicides, and urine – not to mention topsoil – end up in the drink. As the late, brilliant ecologist Paul Shepard put it, "If there is a single complex of events responsible for the deterioration of human health and ecology, agricultural civilization is it. At its worst, agriculture is industrial and corporate, poisoning the whole planet with chemical compounds not found in Nature."[11] Where agriculture is concerned, a revolution in public perception, private stewardship, corporate responsibility, and government control is long overdue.

As the height of land moves west to encompass the Fox and Wolf River drainage basins, we enter Wisconsin, noted for its dairy farms. Here we find lots of effluvia and cow flatulence (a pollutant in itself when emitted on a grand scale) but little in the way of protected areas. Undue enrichment of creeks and rivers has led to major problems in the Green Bay area. Nonetheless, these are being tackled by scientists and citizens' groups, as Hallett Harris and Val Klump indicate elsewhere in this book. Some small state parks and National Wildlife Reserves on the Door Peninsula and offshore islands help to redress the balance. So do some parts of the large Nicolet National Forest to the north. But like the big wildlife reserves in distant Quebec, Nicolet is a multiple-use area, which includes extensive logging and other destructive practices. National and state forests in Michigan's Northern Peninsula are also multi-use areas; nevertheless, when managed intelligently for a wide range of forest values, they afford some headwater protection. In general, this is a far wilder region than the agricultural and urban country to the south and contains important natural areas such as the Seney National Wildlife Reserve, Tahquamenon Falls State Park, the Porcupine Mountains Wilderness State Park, and the magnificent Isle Royale National Park, a haven for moose, wolves, and other wildlife.



At the western tip of Lake Superior, near Duluth, the height of land again approaches the shore before swinging inland to encompass the St Louis, Whiteface, and Cloquet rivers. The Vermillion Range and Sawtooth Mountains mark the edge of the watershed here, and a number of Minnesota state parks such as Tettegouche, Caribou Falls, and Temperance River help to protect headwater rivers. Here, far from the Adirondacks, we are again traversing Precambrian Shield uplands. At the Pigeon River, which in part divides Minnesota from Ontario, the La Verendrye Waterway Park protects what was once the heavily travelled route of fur trade canoe brigades.

Northeast from the old voyageur and Ojibwa rendezvous at Grand Portage, we travel the valleys of the Nor'Westers, a range of hills and mountains near Thunder Bay, Ontario. While we are still in a mixed forest zone of deciduous and coniferous trees, at higher elevations we find advance units of the great boreal forest that stretches far into the North. East of Lac Des Mille Lacs, the divide lies in swampy country once crossed by very long and despised portages: the Portage de la Prairie, the Portage du

Milieu, and the Savannah (Savanne) Portage. Motorists speeding across the Trans Canada Highway today probably don't notice that this is the height of land that divides the Hudson Bay watershed from that of the Great Lakes–St Lawrence.

About two hundred kilometres (125 miles) north of Lac Des Mille Lacs, we approach the place where Paddle to the Sea began his long journey. A decade ago there was little to celebrate along this stretch: much of it was in forest industry hands. However, in 1995 Wabakimi Provincial Park, located above Lake Nipigon (another claimant for the status of sixth Great Lake), was greatly expanded to 900,000 hectares (about 2.2 million acres) to protect woodland caribou habitat. Further expansion in 1999 made it one of the world's greatest boreal forest parks. The Wabakimi area,which is found on both sides of the height of land, has become a model in terms of headwaters protection. Lake Nipigon itself gained greater protection in 1999. Its islands – many of them important caribou calving sites – as well as large spans of mainland shoreline will now be saved from development. The extensive new protected areas around Lake Nipigon are the heart of a waterway and land corridor linking the huge Wabakimi park to Lake Superior.[12]

This success story did not just happen: Ontario has a distinguished group of conservationists and wilderness advocates with on-the-ground experience as well as strong theoretical and tactical knowledge. Determined fighters against corporate greed and government indifference, they have gone into battle many times in the last five decades and suffered many losses; yet they have always returned to the front. Though far removed from the highly paid professional lobbyists of the business and corporate world, they have monuments in the Wabakimis, and Queticos, and Algonquins, and Killarneys, and Temagamis of Ontario, and hundreds of lesser-known sanctuaries.

The most recent of these victories stems from the Ontario government's commitment to represent the province's thirteen natural regions in parks or other conservation areas. This grew out of the World Wildlife Fund Canada Endangered Spaces program, a long, hard, and costly campaign supported by thousands of individuals and hundreds of conservation groups. Behind this in turn were research and recommendations of the World Conservation Union, and the 1987 Brundtland Report of the World Commission on Environment and Development, *Our Common Future*. The report called for protection of at least 12 percent of each nation's land in national parks or equivalent preserves. In Canada the related World Wildlife Fund program took 12 percent as an absolute minimum, recommending 15 to 20 percent in some areas. In 1992 the Endangered Spaces goal was accepted as public policy for all thirteen jurisdictions across Canada: federal, provincial, and territorial. "No nation worldwide has a conservation strategy that can match Canada's in terms of vision, breadth, scientific defensibility, public acceptance, and probability of ultimate success," conservation biologist Reed Noss now observes.[13]



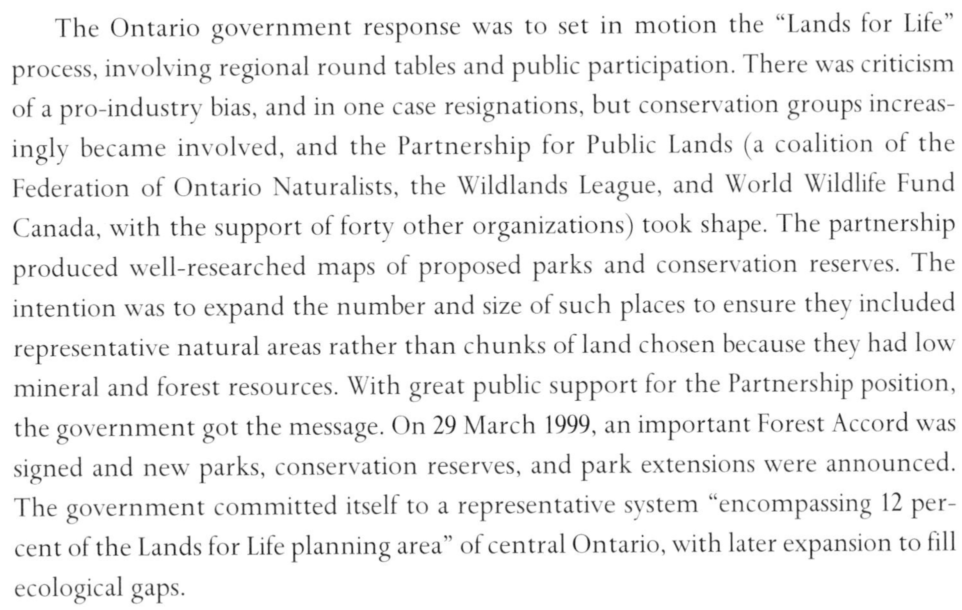

The Ontario government response was to set in motion the "Lands for Life" process, involving regional round tables and public participation. There was criticism of a pro-industry bias, and in one case resignations, but conservation groups increasingly became involved, and the Partnership for Public Lands (a coalition of the Federation of Ontario Naturalists, the Wildlands League, and World Wildlife Fund Canada, with the support of forty other organizations) took shape. The partnership produced well-researched maps of proposed parks and conservation reserves. The intention was to expand the number and size of such places to ensure they included representative natural areas rather than chunks of land chosen because they had low mineral and forest resources. With great public support for the Partnership position, the government got the message. On 29 March 1999, an important Forest Accord was signed and new parks, conservation reserves, and park extensions were announced. The government committed itself to a representative system "encompassing 12 percent of the Lands for Life planning area" of central Ontario, with later expansion to fill ecological gaps.

East of Lake Nipigon the watershed divide bends south toward Lake Superior, passing through both mixed and boreal forests. Conditions are varied, with much heavily cut-

over land and a maze of tote roads, but other areas such as that surrounding the Gravel River have been saved under the Lands for Life process. Extensions have been made to Pukaskwa National Park and Lake Superior Provincial Park, both magnificent places with towering granite cliffs, rushing waters, and rare plant and animal communities. Hiking or canoeing them now, it is easy to forget their relatively recent vintage in terms of protection. Pukaskwa became a provincial Wilderness area in 1964 and fourteen years later was transferred to the federal authorities to provide the basis for a national park.[14] Lake Superior Provincial Park, established in 1944, by the late 1960s had become severely degraded. In 1970, on behalf of the Wildlands League, I inspected portions of it in the company of Douglas Pimlott, Canada's leading ecologist/zoologist/forester/environmentalist, C.B. Cragg, a brilliant York University chemist and president of the Wildlands League, and Dennis Voigt, a forward-looking government forester and avid wilderness traveller. We already knew that approximately 95 percent of the park was under timber licences. We soon discovered what that meant: ruination of the beautiful Sand River Valley, eroding skidder trails and bulldozed roads that blocked trout streams, and liquidation cutting of giant yellow birch and pine in potential nature reserve areas. Such poorly controlled multiple-use practices were enormously destructive and certainly unsustainable. Accordingly, in 1971, Doug Pimlott and I edited *Why Wilderness*, a book that focused on the mismanagement of the park. It had impact, and eventually the heavy logging of Lake Superior Park was reduced and nature reserves established. With the 1999 expansion of the area, commercial logging is out and things are much better.

Following the height of land in a south-easterly direction we encounter large hydro dams on rivers such as the Michipicoten and Montreal. For many decades this Algoma country, spectacular "rock knob" terrain with hills and low mountains rising high above Lake Superior, has been a magnet for talented painters including the Group of Seven. Almost all of it is crown land, clothed with mixed forest of red and white pine, maple, white and yellow birch, spruce, tamarack, and poplar. Particularly in the autumn its flaming colours and grand contours are striking.

A significant section of the Algoma Highlands north of Sault Ste Marie has recently been removed from logging and given protected status. This area east and north of Ranger Lake contains towering old-growth white pine and massive yellow birch. To the east, important park additions have been made to the Mississagi River headwaters, including at Biscotasi Lake, once a favourite haunt of Grey Owl (Archie Belaney). Again, these additions, totalling about 64,500 hectares (159,300 acres), are the result of negotiations during the Lands for Life process.



But these new-blooming flowers should not obscure the fact that not all is lovely in the garden. Further upstream east of Chapleau, in huge forest-industry fiefdoms the clear-cuts go on and on. Denuded areas include those where the Spanish River rises in little streams on their way to the North Channel of Lake Huron. Forest industry giants have cut right across ponds and rivulets that feed the height of land rivers. Without forest cover, there is no retention of soil or water, and no shaded shelter for aquatic life – scant ecological sensitivity, little restraint, and damn all caring. The many large signboards flaunting the reforestation programs seem cynically self-serving.

Should the forest industry firms be cut off? Of course not: they are vital to northern communities and the general economy. But they should be constrained and directed, and should accept that forests are valuable for more than wood chips or board-feet of lumber. There can be no more "cut and get out" approaches that have been damaging to both workers and forests. It is time, too, that the companies ceased to attach blame for job losses to conservation groups: in large degree such losses are caused by technological changes that save the companies money. Another cause is ill-considered cutting cycles that exhaust forest resources, and insufficient reforestation. Where winters are long and the sandy soils thin along the northern height of land, trees do not spring up like beanstalks in (irrigated) Californian valleys.

The streams of the Spanish River headwaters have been abused, but further downstream the 1999 Lands for Life determinations are promising. A waterway park more than 120 kilometres (seventy-five miles) long and with an area of 33,825

hectares (83,550 acres) has been announced. Some you lose, some you win; on balance, for nature 1999 was a winning year in Ontario.



Paddling the fabled Temagami country to the east over a century ago, Archibald Lampman captured its character: "All day we saw the thunder-travelled sky / Purpled with storm in many a trailing tress, / And saw at eve the broken sunset die / in crimson on the silent wilderness."[15] There is still wilderness in Temagami, despite great pressure from would-be resort developers and the forest industries. This much smaller wilderness lives on because of a "thunder-travelled" history of conservation research and activism in which the Teme-augama Anishnabai, the Algonkian-speaking "people of the deep water," have played an important role. The Maple Mountain notion of creating a major recreational complex and resort in the core of the Lady Evelyn Wilderness – with golf course, tennis courts, and accommodation for 3,500 people – was strongly opposed by Temagami Chief Gary Potts. He was joined by Hugh Stewart who headed up the Save Maple Mountain Committee, but it was primarily legal cautions filed by Potts and his colleagues that stopped the mega-development.

Temagami, with the Lady Evelyn Smoothwater park at its core, endures because of hard work on the part of the Teme-augama Anishnabai, the Alliance for the Lady Evelyn Wilderness, the Save Temagami Committee, the Temagami Wilderness Society, and associated groups. In 1989 the Wilderness Society even enlisted the support of Bob Rae, then opposition leader in Ontario's Legislature, who joined environmentalists blocking the Red Squirrel logging road extension and was arrested for his trouble. At the same time, a forest research program supervised by Dr Peter Quinby established Temagami's importance as "the last significant old-growth pine ecosystem in North America."[16] This led, rather grudgingly, to increased forest protection in this important headwaters arena – long an area of disgraceful mismanagement on the part of the Ministry of Natural Resources. As Gordon L. Baskerville, dean of forestry at the University of New Brunswick, put it in 1986, "Much of the planning material in this area would be better described as creative writing about the resource than as a realistic attempt to control resource development over time to achieve objectively stated values."[17] The plans, put simply, were designed to accommodate local timber companies.

During 1998, the Partnership for Public Lands recommended major additions to Temagami's Lady Evelyn Smoothwater Wilderness Park. Several conservation reserves close to but unfortunately not contiguous with the park were added, as were several additional reserves near the provincial border with Quebec north of Lake Timiskaming.

The magnitude of what was achieved in 1999 is still sinking in and raising discussion. The Lands for Life/Partnership for Public Lands effort increased the amount of protected area in Ontario by 2.4 million hectares (about six million acres) – an expansion greater than the total area of the State of Vermont!



While I am tempted to exit Ontario on this triumphant note, it must be added that Lands for Life was not perfect. The government may allow mining in some park areas, and hunting has been enshrined in most new parks and introduced as a possibility even in existing wilderness parks. The concept of sanctuaries for living things has yet to illuminate the minds of our political leaders. But as Ric Symmes, executive director for the Federation of Ontario Naturalists, wrote, "We cannot wait for perfection to start protection, and the critical need is to secure the parks from heavy industrial use, while there is something left to protect."[18]

As we follow the divide back into the vast province of Quebec, no longer do the rivers flow into the Great Lakes. Their destination now is the Ottawa River or Fleuve Saint-Laurent. To be blunt, Quebec has not done the necessary job where headwaters protection is concerned. We began our journey around the great divide that marks the outer limit of the Great Lakes–St Lawrence River watershed in Quebec, south of the St Lawrence. In that area, it was noted, only 1 percent of the land was truly preserved. In 1995, the percentage for the entire province was far worse, with only 0.4 percent of Quebec legally protected to Endangered Spaces standards. Among other things, those standards call for undeveloped, roadless, essentially wild areas of significant size. However, the province has yet to establish enough of them, even though it committed itself to complete a system of representative protected areas by the year 2000. At the turn of the millennium, the political promise seems hollow, although modest proposed additions bring the total protected area to 4.2 percent.[19] Again, a glance at a map may appear to contradict such assertions: after all, the large Reserve faunique de la Verendrye is shown all in splendid green. A closer look reveals that it is riddled by highways and roads, many for hauling out timber. World Wildlife Fund Canada sums up the situation: "There has not been a wilderness park established in Quebec for over 10 years."[20]



Mount Tremblant Park, while considerably south of the height of land, is acclaimed even by conservationists as a truly protected place. Within its boundaries lie four hundred lakes, seven rivers, and countless fast-flowing streams. But farther north along the divide that tips Quebec's marvellous rivers south, there is virtually no legal, formalized protection. Some of these rivers are mighty, but few are free-flowing. On many of them, Paddle to the Sea would be blocked by giant dams. Hydro-Quebec, which is central to the economic and political life of La Belle Province, has harnessed and diminished its once-wild waterways on a scale reminiscent of the U.S. Army Corps of Engineers in more unrestrained days. It is a century since Shawinigan Falls on the St Maurice River became the first of the major hydroelectric complexes in Quebec. Many more followed: the Saguenay, the Bersimis, the Ottawa, and the massive Manicouagan-Outardes Project. This latter structure, about 230 metres (750 feet) high, contains an impoundment five times larger than Lake Mead at the Hoover dam in the U.S.[21] In light of the James Bay megaproject, however, these were just warm-up exercises for Hydro-Quebec.

Wild, clean streams still tumble off the Laurentian Plateau, but they become fewer in number as hydro dams designed to feed the international power grid proliferate. The environmental impact seems to have been of little interest to the late Premier Robert Bourassa, who wrote in 1985: "Because of its particular climate and topography, Quebec is a vast hydroelectric plant in-the-bud, and every day millions of potential kilowatt-hours flow downhill and out to sea. What a waste!"[22] The ramifications of this attitude can be seen in destroyed waterways, scars on the land, drowned valleys, displaced fish and wildlife, and disrupted native communities.

Yet there are still some sun-pierced silences in Quebec, places that are protected to provide sanctuary for wildlife and for the enjoyment of future generations. Some, such as Parc national de la Mauricie, Parc de la Jacques-Cartier, and Parc des Grands-Jardins, are far south of the height of land. In the north of Quebec, the spruce-fir forest gives way to the "land of the little sticks" – the taiga, a landscape of scattered, stunted trees and shrubs. There is much wild land here, but it is open to industrial exploitation until formally protected. From these heights of precambrian rock, truly majestic rivers flow south: the Moisie, the Ste-Marguerite, the Saguenay and its northern extensions, and the Magpie. We can hope that at least some of these will be protected from the insatiable appetite of Hydro-Quebec: several of the largest have already been scheduled for damming.

One valuable area far south of the height of land has recently been spared. This is the very scenic upper gorge of the Malbaie River where a 22,550 hectare (55,600 acre) park has been established. But Quebec still has a long way to go, despite the efforts of many hard-working environmental groups.[23] At the moment, large wildland preserves are being urged by organizations such as l'Union québecoise pour la conservation de la nature. But with industrial development rapidly spreading, time is running out to secure such areas.

It is a long journey around the edge of the Great Lakes–St Lawrence River watershed. Along the divide we have encountered everything from striking, sparsely settled mountains to crowded lowlands occupied by great cities. Yet this is all height of land country, whether it be Chicago's south side or the lonely Moisie River headwaters in Quebec, the barely perceptible divide at New York's polluted Erie Canal or the shield wilderness of Ontario's Wabakimi Park. Bedrock, soils, vegetation, climate, elevations, and local drainage patterns vary enormously along our route, resulting in markedly different settlement patterns and human activities. The focus here, however, is on healthy biological diversity, on protected areas, and on a clean environment. The International Union for the Conservation of Nature and the Brundtland Report recommend that 12 percent of the earth's surface, at a minimum, be set aside to conserve and protect species and ecosystems. In the huge Great Lakes–St Lawrence watershed, we are far from that goal.

Natural integrity and headwaters protection have too often been ignored on our journey to the sea. This is so even along the outer boundary of the watershed. Downstream, the problems are compounded, but the highland headwaters are where it all begins. If the vital rivers that spill into the Great Lakes and St Lawrence are to thrive and deliver health and life to the lakes, to plants and animals (including the human animal), they need a good start in life. At the very least, they must be spared abuse. Paddle to the Sea encountered problems downstream, but he had a strong beginning in his clean northern stream. Perhaps that is why he survived his long voyage. Our watershed rivers need the same, and we need them.



Facing page: Frontenac Provincial Park, Ontario. Photo: Gregor G. Beck

# Sane Friends of the Mad River

## *Downstream from Vermont*

*Elliott Gimble*

*"Away from the lively meetings and consensus-building, the trees sway and the birds sing, oblivious to ongoing human activities to secure the watershed's future ... By creating a shared vision and addressing conflicting uses, the Mad River's human populations can learn together to safe-guard this watershed's vital resources for the next generation and beyond."*

ABOUT 225 KM (140 MILES) southeast of Montreal, Vermont's Route 100 cuts through a narrow, wooded reserve called the Granville Gulf State Scenic Area. Tourists and residents alike are attracted by spectacular autumn colours and picturesque views along this two-lane highway. Those who take the time to stop can sense nature's power in Granville Gulf: the wind rustling leaves, the small but steady work of water gurgling over rock. Signs of human development seem distant.

Few probably realize that they stand along the edge of a watershed, that ridge of land separating distinct drainage basins. Rain falling on the south side of Granville Gulf forms the headwater streams of the White River, which in turn flow south to the Connecticut River and ultimately to Long Island Sound. On the other side, a small mountain brook, narrow enough to step over, marks the source of the Mad River. Flowing north for 40 km (25 miles), it feeds into the Winooski River in Moretown, due west of Montpelier, Vermont's capital. In turn, the Winooski makes its way into

Facing page: Mad River, Vermont. Photo: Gregor G. Beck

Lake Champlain, which flows north into the Richelieu River and finally to the mighty St Lawrence.

For the 600,000 people who live downstream in the Lake Champlain watershed, what happens in headwater regions like the Mad River Valley is of vital importance. The potential impacts of headwater policy range from economic to public health to purely environmental. These are no small issues for the Champlain Valley, a region that reaps $2 billion annually from tourism, draws one-third of its drinking water from Lake Champlain, and is home to more than eighty fish species and over three hundred bird species.



The peaceful, pristine nature of Granville Gulf belies an extensive human history in the Mad River Valley. After settlement, the watershed had its share of trapping, logging, and farming. Fifty-one mills during the 1800s processed area lumber into clothespins, doors, bowls, and the like. Logging gave way to sheep farming and, later, dairy farming. Three ski areas opened in the seventeen years following World War II, heralding growth in population and recreation-tourism businesses.

Today, forests cover about 86 percent of the Mad River watershed. In addition to the ski developments, the valley still supports active farms with good agricultural soils and spectacular scenery. Its headwaters maintain excellent trout populations, and the river and its tributaries are home to some of Vermont's best swimming holes.

But the Mad River has its share of problems too. Inadequate erosion-control measures, poor management practices, and a lack of stream-bank vegetation have accelerated sedimentation, fouling the habitat of aquatic organisms and raising water temperatures. Failing septic systems leak high levels of fecal coliform, and storm runoff carries other pollution into the river, creating accelerated algal growth and health risks for swimmers. Water withdrawals (mostly for ski resort snowmaking), gravel removal, and other human activities are altering stream dynamics and fish habitat and affecting recreational uses.

Fortunately, the Mad River has its friends. Since 1985 a group of grassroots citizen volunteers under the Mad River Watch Program has regularly gathered water-quality data on the river, testing for pH, temperature, and turbidity, sampling aquatic insects and invertebrates, and recording physical characteristics. In 1990 representatives of towns, planning commissions, and regional conservation organizations formed the Friends of Mad River, "a non-profit organization dedicated to protecting and improving the ecological, scenic, and recreational values of the Mad River and its tributaries."

In November 1993 the Friends of Mad River and the Mad River Valley Planning District received a grant from the United States Environmental Protection Agency to develop a demonstration project that would identify the river's uses and assets while determining solutions to its various threats. A collaborative process began, with participation by landowners, government agencies, businesses, community organizations, and residents. During 1994, project subcommittees gathered information into topic papers and met with town-planning commissions. A public outreach subcommittee organized open public forums in watershed towns, and more than one hundred residents participated, sharing their visions for the Mad River's future. A river walk brought together residents and river managers on site to discuss river dynamics and issues.

The final Mad River Conservation Plan, including results of the town meetings, topic papers on various issues, and maps, was finalized after a public comment period and was distributed in early 1995. The plan's recommendations address all stakeholders, calling for towns to revise plans, zoning ordinances, and regulations to provide for adequate stream buffers; schools to develop a comprehensive river curriculum; non-profit volunteer groups to develop a river path and to train landowners in stream inventory; businesses to consider cumulative effects of their activities in the watershed; governments to set up water-quality sampling sites, provide technical and financial assistance to farmers, and plan for fish-habitat restoration; and, individuals to support river-watch programs and organizations and to take part in an Adopt-a-River program.

As this sampling of recommendations demonstrates, all those who depend on and care for the Mad River have a vision to share and a recognized role to play in protecting their home watershed.

Up in Granville Gulf, away from the lively meetings and consensus-building, the trees sway and the birds sing, oblivious to these ongoing human activities to secure the watershed's future. The river Conservation Plan makes the connection – that by creating a shared vision and addressing conflicting uses, the Mad River's human populations can learn together to safeguard this watershed's vital resources for the next generation and beyond.

# Over the Gunnels in Algonquin

## *Clean Waters – Straight from the Source*

*Jeff Miller*

OF THE TWO THOUSAND LAKES IN ALGONQUIN PARK, the one most appropriately named is Source Lake. In fact, "Source" could easily be an alternate name for Algonquin, Ontario's premier provincial park. This three-thousand square mile undulating dome is the starting point of a dozen small rivers, navigable by canoe and kayak, which flow into the Great Lakes and St Lawrence River system.

It is here, some 150 miles north of Lake Ontario, that the Madawaska gets serious and starts. It gets spunky right away, then loafs around in a moose and beaver swamp before chugging and wiggling through a long valley of its own making, to the south and mostly east. Eventually, if you were pollen or lichen dust or a long-distance canoe tripper, you could go all the way to the Atlantic Ocean with assistance provided by the Ottawa and St Lawrence. On the western side of the dome, it is a fight all the way for the Magnetawan. The waters crash and tumble toward the wind-laced pines of Georgian Bay, and then basin out for a while in Huron, Erie, and Ontario. Soon after the confusion of the Thousand Islands, they mix again with the Ottawa supply and shoot out big into the St Lawrence River valley.

Lots of folks along the way of these bumpy rivers count on this pure Algonquin water – contaminated mostly by mink pee, moose pellets, and bear sweat. Surely these romping young rivers mothered in the park mean living for a hell of a lot of ecosystem between Algonquin, the big lakes, the river and the sea. And so we pray for the water, and still we get it, mink pee notwithstanding.

For fifty years I have gone over the gunnels in these waterways, and their names are soaked into my DNA – Bonnechere, Opeongo, Amable du Fond, Petawawa. In the winter, with your ear to the frozen Petawawa, you will hear it below, rumbling along in its rocky conduit – a secret of the snows. You cannot hear it at all where the river widens to lake, as under several feet of blue ice it rolls along like oil, black, timeless, immense. At the dim bottom of Lakes Lavielle, Merchant, and Lamuir, the big trout doze until spring's arrival connects them once again to the sky.



Spring, summer, and fall bring welcome change, investing the waters with golds and greens, and a cerulean blue which can lose control of the sky and paint the water purple. The giving goes on. Oxygen from the massive mixed forest combines with the watery molecules in their frolic from the steps of the park to nourish life downstream.

Yes, I know well the ways of Algonquin's waters. By the gallon, they have thinned my acrylics, mixed my watercolours, and endowed me with multitudes of wet ideas for my canvases. Algonquin insists that I do this full time – and so I do.

When I was a gangly kid, Algonquin's linkages of lakes and streams still carried the big softwood on a journey through hundreds of white pine chutes, and past dams and camboose camp remains. A century earlier, the long, square-hewn sticks without a knot went to England, so that her navy could be assured of ship masts from within the empire. The water provided reliably, as the cathedral pine was removed forever. Today in Algonquin Park, on the dome where the rivers start, industrial logging goes on. It is a different, mechanical logging in a different kind of forest. It operates under the scary mandate of "sustainable development." Will the water apparatus of Algonquin continue to provide so generously? Tomorrow's answer may lie only in the hope that rides uncomfortably in the guts of those of us who know nature for our home.

Art fosters an environmental ethic. Photo: Gregor G. Beck

When forests are cleared right to the edge of streams, rivers, and lakes, there is greater risk of flash floods and increased water pollution from pesticides and herbicides. In agricultural areas, excess fertilizers and manure end up in the water as well. Forest buffers along the water's edge reduce these environmental problems, provide critical habitat, and serve as natural corridors for wildlife.

Photo: Phillip Norton

# Getting Our Act Together

## *The Greening of the Châteauguay River Watershed*

*Serge Bourdon and Phillip Norton*

*"Unfortunately, it is the collective damage of hundreds of small farms throughout the watershed that is wreaking havoc on a natural resource that should be of benefit to everyone, and to the river. And – just like industries, municipalities, and households – farmers, if they are not pushed a bit, will not change."*

From its headwaters in the Adirondack Mountains of upstate New York, the Châteauguay River begins as a broad network of tributaries which wind their way north into Canada. Wild, rushing trout streams and the spectacular 37 metre (120 foot) High Falls on the U.S. side gradually yield to a wider, slower river in Quebec.

The Châteauguay's historic main branch passes beneath the Powerscourt covered bridge (built in 1861), past the Battle of the Châteauguay National Historic Site (War of 1812), past some of the richest dairy farms in the province, and through small towns and parishes which bestow upon the river a unique blend of French and English culture – Huntingdon, Très St Sacrement, Sainte-Martine, Ormstown, and Allan's Corners among others.

In Quebec the Châteauguay turns northeast and flows parallel to the St Lawrence River as if in a test of might with its inevitable master. One hundred and twenty kilometres (75 miles) from its wilderness origins, it is tamed into a suburban watercourse, bordered by exclusive, riverside homes, small industry, marinas, and cement retaining walls. Another seven kilometres (four miles) past its namesake city of over 42,000, it touches the Mohawk Reserve of Kahnawake before merging with "le fleuve St-Laurent."

While the Châteauguay River is unusual in its international, multilingual, and multicultural aspects, and its unusual geological and human history, it is also quite typical of the scores of St Lawrence tributaries in "la belle province."

These rivers, flowing through the long-settled agricultural bottomlands of the St Lawrence, face similar threats to their water quality – namely, pollution from municipal and industrial sources and agricultural wastes. For decades the Châteauguay River has been the servant to a few, to the detriment of many. Toxic chemicals from industry and raw sewage from towns accounted for half of the problem, with "non-point" sources of pollution accounting for the rest.

Agricultural seigneuries of New France were established here as early as 1673, but only now have the effluents of human habitation and industry begun to be treated. Thanks to Quebec government subsidies and local citizen pressure, six of the seven towns directly touching the river have new wastewater treatment plants. Only one stand-out village has been unable to commit to such a project, claiming that the costs of building and maintaining a plant would surpass the financial means of its 659 citizens. While there are also a few small hamlets without proper wastewater treatment, any new home construction must conform to municipal regulations for septic systems.

Quebec's investment in municipal wastewater treatment is a great accomplishment. Yet Denis Brouillette, a biologist for the ministère de l'Environnement et de la Faune (MEF), recently stated that the next challenge – that of cleaning up each of the non-point sources – will be even greater. The MEF in 1996, in cooperation with a local watershed group called SCABRIC, produced a brochure entitled "The State of the Aquatic Ecosystem of the Châteauguay River and Its Watershed, 1979–1994."

In such a tightly knit rural community, pointing fingers at polluters is a very touchy issue. Farmers who inherited the old ways of cattle grazing, ploughing, and manure spreading do not believe that they should have to foot the entire bill for installing proper riverside fencing, pumps, and manure-storage pits. Even the least expensive method of field stacking of manure on a concrete slab could cost upwards of $70,000 for a one-hundred cow operation. This compares to $15,000 annually for the daily hauling and spreading of manure. However, spreading manure on frozen fields is illegal in Quebec because the enriched spring meltwater tends to run off directly into ditches, creating a toxic shock to aquatic life downstream.

In occasional Environment ministry "sting operations," several prominent Châteauguay Valley family farms have been fined $1,000 for winter manure spreading. While no one is denying that laws are simply being enforced, the agricultural community says that the government is fining the easy-to-get little polluters while the big guys are left alone.



Unfortunately, it is the collective damage of hundreds of small farms throughout the watershed that is wreaking havoc on a natural resource that should be of benefit to everyone, and to the river. And – just like industries, municipalities, and households – farmers, if they are not pushed a bit, will not change.

Understanding the farmers' point of view is the first step to finding an acceptable solution to this thorny problem. Around the council table at the Sainte-Martine town hall, a diverse group of valley citizens meets each month to hash out such issues. They are farmers, mayors, business people, educators, tourism promoters, and environmentalists – all board members of Quebec's first watershed management agency. La Société de conservation et d'aménagement du bassin de la rivière Châteauguay (SCABRIC) was officially chartered in 1993, the culmination of a nearly decade-long campaign by outdoor enthusiasts and ecologists to clean up the river.

Corresponding with the international wave of renewed environmentalism in 1990, this grassroots movement raised awareness through the media, lobbied government representatives, and carried out projects to plant trees and prevent soil erosion. However, publicity about agricultural pesticides, fertilizers, and manure often brought confrontation, and ecologists soon realized that nothing was going to change if one side was pitted against the other.

Through SCABRIC, a clear and simple common objective has been stated – to clean up the river – and priorities have been set on how to achieve this with the least pain. Environmentalists are understanding the financial concerns of farmers, while farmers are realizing that cleaning up can actually be financially beneficial in the long term. Proper manure storage can yield reduced fertilizer costs and improved soil conditions, in addition to a healthier environment on the family farm.

Sylvie Laberge is a dairy producer from St-Paul-de-Châteauguay, Quebec, and a former director on the SCABRIC board: "We are not a bunch of polluters," she says. "We are of the young mentality that is trying to keep pesticide use to a minimum and to practise sustainable agriculture."



The regional branch of l'Union des producteurs agricoles farmers' union held a conference in Valleyfield in 1995 which was in tune with the new environmental theme "I take care of my earth." Farmers are not only turning "green" because it is good for the environment but also because it can pay off. There are positive signs of change in the way the new generation is thinking, a willingness to look at their part of the problem and to take measures to correct it.

Placing a dollar figure on the value of watershed management is often difficult – except when it comes to flood prevention. The City of Châteauguay, Quebec, made national news in Canada when a January 1996 thaw and ice jam sent dirty river waters into the streets and into 690 homes. In Châteauguay alone, damages were estimated at three-quarters of a million dollars, plus an additional one-half million dollars in costs to the city.

As the year proceeded, Châteauguay Valley residents were victims to two more floods, this time caused by heavy rains in July and November. Flash flooding would not occur under natural conditions since forests and wetlands slow the runoff; but in the name of housing and agriculture, we have cut down our forests and drained our wetlands so that the rainfall which strikes bare fields and pavement flows immediately to swollen watercourses. Two lives were lost in the July torrent, and homeowners once again suffered thousands of dollars in property damages. The parallel to major floods in other watersheds, such as the Saguenay, Red, and Mississippi Rivers, is striking.

Since the first small gathering a decade ago of an unusual mix of "river people" – a fisherman, a journalist, an educator, and several clergy – which attacked the seemingly insurmountable challenge of bringing the Châteauguay River back to health, significant advances have been made. Awareness has been raised at all levels, including improved communications with the upriver American neighbours. Despite the French-English language barrier and political separateness, SCABRIC has been important in opening a dialogue between Quebec's Environment ministry and its New York counterpart. Schools on both sides of the border are forming Internet alliances to share data on river water analyses. And, since 1993, an annual international forum, alternately held in the United States and Canada, has featured speakers and workshops on all aspects of the watershed environment, with headphones for simultaneous translation.



The river is now regarded as an integral part of the entire *watershed* which transcends all human boundaries and must be cared for through regional and international cooperation. A watershed map has been published, a resource library has been started, trees and flowers have been planted, an educational slide show is touring the schools, and conferences have established a network among Americans and Canadians, French and English.

A master plan is in place stating specific challenges that lie ahead. Case by case, solutions must be found for eliminating soil erosion, minimizing flooding, fencing cattle away from the river, keeping manure on the farm, correcting leaking septic tanks, cleaning up automobile scrap yards, leaving unploughed grass or tree buffers between field and streams, and teaching households, schools, and businesses to dispose of toxic wastes properly.

Yes, it is a dream; but those who share this dream are not sleeping. The core of enlightened citizens that is SCABRIC believes that their vision of watershed management will yield long-term benefits far outweighing the short-term costs.

# Fouling up on the Farm

## *Agricultural Non-Point Sources of Pollution*

*Hallett J. Harris and Val Klump*

*"Present erosion rates are not only an economic drain but are clearly non-sustainable. It takes between two hundred and a thousand years to develop just one inch of topsoil on croplands. Yet in the United States, one inch of soil is lost every 16.5 years."*

OVER 10,000 YEARS AGO, as the earth warmed and the great glaciers receded, meltwaters raced across the landscape through forest, meadow, valley, and wetland. Mere trickles at first, they coalesced to form brooks, streams, and rivers that filled the lakes of the most spectacular freshwater system in the world: the Great Lakes–St Lawrence watershed. They traced upon the land an intricate network that still feeds these lakes and rivers today, like the complex, finely branching root systems of a mighty oak or pine. And like the roots of a tree, this network of rivulets to rivers steadily nourishes the lakes with nutrients and minerals weathered from the rocks and soils of the hinterlands and the shorelines. These nutrients supply the basic building blocks of life to the food web of the lakes, from microscopic algae to trophy-sized trout.

In perhaps as few as ten generations, we have accelerated the transport of materials from land to lakes a thousand-fold or more. Forests were cleared, meadows were put to the plough, and towns and great cities arose. Today, the natural flux of nutrients and soils delivered by this same network of

Facing page: Livestock with uncontrolled access to the water's edge cause numerous environmental problems. Manure and urine pollute the water with excess nutrients and can cause major health risks. Crushed vegetation and trampled shorelines also lead to faster soil erosion. Photo: Gregor G. Beck

watercourses is often swamped by rapidly eroding soil particles, excessive nutrients from nitrogen and phosphorous fertilizers, and runoff from farms and city streets. These pollutants are truly "resources out of place."

Hardest hit by this new onslaught of material flowing downstream are our shallow coastal embayments, river mouths, and harbours. By virtue of their immense size, the open Great Lakes (especially the deep upper Great Lakes) can dilute and assimilate significant inputs of nutrients. This is not the case, however, in coastal embayments like Green Bay and Saginaw Bay, or in Lake Erie or the small lakes along the St Lawrence. Hence, chronic nutrient enrichment occurs.



Soil particles and plant nutrients have two primary effects on water quality: the loss of water clarity and the stimulation of aquatic plant and algae growth. This over-enrichment causes "eutrophication." Eutrophic waters have profuse and excessive plant growth fuelled by an unlimited supply of nutrients. When, in the natural course of events, these plants decay through bacterial action, oxygen in the water is consumed. Fish, which need the oxygen dissolved in water, suffocate and die. Other resultant problems include turbidity, foul odours, tainting of water supplies by blue-green algae species, and unsightly algae blooms. Less obvious are changes in species composition of biological communities. Diverse, native populations of organisms are gradually displaced by a few pollution-tolerant inhabitants capable of surviving in oxygen-depleted waters.

The increased turbidity or cloudiness prevents light from reaching the bottom, and underwater plants can no longer survive. This rooted, aquatic plant community provides an essential habitat for aquatic invertebrates like insects and crustaceans, which serve as the food source for small fishes such as minnows, which in turn feed the larger fishes and aquatic birds. The end result is a drastically altered ecosystem. Biodiversity has been reduced, and many of the desirable qualities of the lake have been lost.

These pollutants are classic examples of what is termed "non-point source" pollution. Non-point source pollution enters the environment from diffuse, widespread sources, in contrast to "point sources," which originate from discrete, easily identifiable locations. Sources of non-point pollution vary, but agricultural sources (notably soil erosion and fertilizers) are the most pervasive today. If the watershed is to be protected from these pollutants, attention must be drawn to the sources – the lands that surround the lakes and rivers and streams.

Although erosion rates on cropland are not a direct measure of the sediment reaching streams, they reflect lesions on the land that deserve our attention. Soil erosion is influenced by several complex factors, including rainfall, soil erodability, slope steepness and length, type of crop or forest cover, and crop and livestock management. Sediments from erosion are, by volume, the greatest single pollutant of surface waters and are also the principal carrier for chemical pollutants. Although classified as having a non-point source, most of a watershed's sediments may come from a few relatively small areas that need special attention. The soil erosion rates from croplands in the United States average about seventeen metric tonnes per hectare per year (7.5 tons/acre/year), of which ten tonnes/ha is water-borne and the remainder is wind-borne. For comparison, erosion rates in undisturbed forests range from only 0.004 to 0.05 tonnes/ha per year.

The relationship between land use, erosion rates, tributary sediment loading, and water quality is abundantly clear. Of the Great Lakes, Lake Erie has the largest proportion of land devoted to agriculture (67 percent), the highest shoreline use, and the most rapidly growing human population. Despite its small size, it also has by far the largest tributary loading of suspended solids: 6.5 million metric tonnes per year. In comparison Lake Ontario has 1.6 million, Lake Superior 1.4 million, Lake Huron 1.1 million, and Lake Michigan 0.7 million tonnes per year.

Direct and indirect costs, amounting to billions of dollars each year, accompany non-point source inputs of sediment and associated pollutants. Cost estimates for the United States due to water erosion damage is estimated at approximately $7 billion annually. This includes both in-stream damages (affecting, for example, water storage facilities, navigation, and recreation) and off-stream damages including floods, and



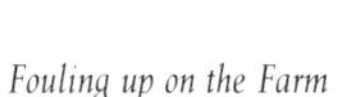

affecting water conveyance facilities and water treatment facilities; more than one-third of this cost is related to recreational damages. A staggering additional yearly loss of $27 billion is attributed to reduced soil productivity. Neither of these cost estimates, however, quantifies the innumerable ecological impacts of lost habitat and biodiversity, or increased susceptibility to disturbance from storms, droughts, or floods.

Present erosion rates are not only an economic drain but are clearly non-sustainable. Soil is being lost far faster than it is being formed. Soils form naturally, but extremely slowly. It takes between two hundred and a thousand years to develop just one inch of topsoil on croplands. Yet, in the United States, one inch of soil is lost every 16.5 years – seventeen times faster than the rate of formation.



Since the late 1940s, the emphasis in agriculture has been on increasing production. To a large extent this has been accomplished by the intensive use of chemicals, including phosphorous-containing fertilizers. Attempts to reduce phosphorous loading in the late 1970s and 1980s focused on point-source abatement, primarily from municipal sewage-treatment plants. This effort has accomplished a significant reduction in loading. However, the more difficult problems of implementation of non-point source reduction remain. Agricultural sources still account for about two-thirds of the present loading to the Great Lakes system.

Management efforts must be directed at both soil and phosphorous loss from the land. Of all the soil-erosion reducing practices applied over the last fifty years, conservation tillage systems are deemed the most efficient methods of dramatically reducing sediment loss while also maintaining productivity. A conservation tillage and planting system is any practice that retains at least 30 percent of crop remains (or residues) on the soil surface after planting.

Interestingly, only 19 percent of the cropland in three of the U.S. Great Lakes states, Michigan, Minnesota, and Wisconsin, were under some form of conservation tillage in the late 1980s. This compares to a national average of 31 percent. Why so little when it is proven to be so effective? The answer is complex, but the bottom line is this – it is not profitable in the short run. With our present economic accounting system, on-site and off-site damages are not subtracted from gains.

Other methods are being used to better manage what ultimately enters and leaves the farm. One approach creates an input/output situation similar to a bank account. What has been learned? Farmers on the average are using far too much fertilizer. For phosphorous, current use is approximately six times more than is needed to keep up production. Consequently, phosphorous must be building in the soil or being lost to other systems. Both are probably true, but again it depends on individual erosion variables as well as the practice of manure handling, fertilizer application, cropping, grazing, stream and shoreline buffer zones, and other factors.

Our present untenable position occurs because individuals in the watershed are making economically rational but ecologically unsustainable choices. While it is possible to make substantial improvements, the reality of the situation is that, for the most part, they have not happened. Perhaps Aldo Leopold was right when he recorded his thoughts in the foreword to *A Sand County Almanac* in 1948, musing that "Conservation is getting nowhere because it is incompatible with our Abrahamic concept of land. We abuse land because we regard it as a commodity belonging to us. When we see land as a community to which we belong, we may begin to use it with love and respect. There is no other way for land to survive the impact of mechanized man, nor for us to reap from it the aesthetic harvest it is capable, under science, of contributing to culture." Have we learned anything more since 1948?

The real Holland Marsh was a vast wetland ecosystem, beautifully diverse with expanses of fen, swamp, and cattail marsh, and home to plentiful populations of fish and wildlife. Over the last hundred years or so, it has been transformed into an area of intensive agricultural production.

Photo: Gregor G. Beck

# Wetland Lost – Farmland Claimed

## *Ontario's Holland Marsh Experiment*

*J. Douglas Blakey and Bruce Litteljohn*

*"The fascinating cornucopia of life forms has been replaced by expansive fields of carrots and onions and celery – dusted with fertilizers, herbicides, and pesticides to produce the perfect-looking vegetables demanded by consumers who may not understand the environmental costs of the unblemished vegetables."*

As rivers go, the Holland is not spectacular. Its main branch rises near the town of Schomberg, less than an hour's drive north of downtown Toronto. From Schomberg the river snakes its way east and north, flowing sluggishly through a broad valley for some forty km (twenty-five miles) to arrive at Cook's Bay, the southernmost extremity of Lake Simcoe. From there its waters follow part of Champlain's exploratory canoe voyage of 1615 and then work their way through the lakes and the St Lawrence to the sea.

Compared to the St Lawrence, the Holland looks like a dull, meandering creek. In 1825 the Scottish novelist John Galt, who was interested in locating settlement lands for the Canada Company, referred to it as "a mere ditch swarming with bullfrogs and water snakes." Given the time and Galt's pragmatic purpose, his demeaning remark is understandable. In fact, what he was looking at in the valley of the Holland River was an 8,000 ha (20,000 acre) biological treasure. This was the *real* Holland Marsh, a huge expanse of peat-rich fen, marsh, and swamp lands about thirty-five kilometres (twenty miles) long by two to three km (1.5 to two miles) wide, bisected by the river itself. Here, for mile after mile, marsh grasses and cattails rooted in spongy peat muck as much as twelve metres (forty feet) deep waved in the breeze. Here, too, sedges grew in wet meadows, while aquatic flowers brightened the scene. From early photographs it appears that much of the area was swampland rather than true marsh (which often has open water, sometimes as deep as one to two metres or three to six feet). Swamps, on the other hand, have much less open water and are dominated by trees and shrubs such as cedar, and hardwoods, along with alder, dogwood, and willow thickets. These grow with "wet feet" or on ground where the water table is barely below the level of the land. Fens, which are thickly underlain by peat and have open water at their edges or centres, were also part of the land and waterscape. The Holland "Marsh," then, was a combination of marsh, swamp, and fen – a tremendously productive natural system providing shelter and food to an enormous variety of wildlife: turtles, deer, loons, ospreys, mink and muskrats, paddle ducks such as mallards and diving ducks such as goldeneyes, migrating geese, beaver and largemouth bass, kingfishers and red-winged blackbirds, bitterns and herons, hawks and owls, rabbits and racoons, weasels and foxes, wolves, black bears and wood ducks. The list goes on and on, for such large and complex wetlands are pinnacles of biological diversity and productivity. The Holland Marsh prior to the end of the nineteenth century was a vast ecological treasure and not "a mere ditch swarming with bullfrogs and water snakes." But then singer Joni Mitchell, in a lyric particularly apt for disconnected lovers and concerned conservationists, has noted that "you don't know what you got 'til it's gone."

The Holland Marsh started to go about 1880. After a few futile attempts to extract peat for fuel or to farm patches of it, thought turned to having a good night's sleep. There was a strong demand for the marsh's sweet wild hay to stuff mattresses. But first there was more work to do: men with scythes went into the marsh and began cutting reeds and long-stemmed grasses. Soon horses with boards strapped to their hooves to keep them from sinking into the muck were pulling mowers through the marsh grasses. By about 1915 mattress-stuffing hay was being cut on about 5,000 ha (12,000 acres) of the Holland Marsh, about half its total area. The people of Toronto may have slept well, but the wildlife of the marsh must have had nightmares – if the creatures even survived the utilitarian onslaught, which in many cases is doubtful.

By 1915 hay cutting had reached its peak. New directions were sought. William Henry Day, a professor at Guelph's Ontario Agricultural College, was invited to look into the draining of the marsh. With much of the land scalped by the mowers, it seemed time to open its arteries and drain the life-blood of this extraordinary area. Well, why not? In the opening years of the twentieth century there were still lots of forest and swamp (even though it had declined substantially from the time of Confederation in 1867). By and large, wild lands were still seen as areas to be conquered and turned to practical account, even though some long-sighted Canadians were beginning to urge conservation measures of a moderate sort.



Professor Day was a persistent man and, it seems, a very bright fellow. He concluded that a system of canals, dikes, dams, and pumps would successfully drain a major portion of the Holland Marsh. His interest was not entirely academic, for he soon got involved in planting vegetables himself and helped to form a development syndicate that purchased 1,600 hectares (4,000 acres) – at about $1 per acre. By 1923 he had resigned his post as head of physics at Guelph and moved to Bradford to push the issue of reclamation of the marsh lands. That meant reclaiming the enormously rich wildlife area for farming. Ironically, "reclamation" today often means returning lands and waters to their natural state. Times and conditions and sensibilities do change.

In 1925 the real task began. Five years later, despite numerous technical difficulties, the work was essentially finished. Much of the marsh was arable. The price of the land had risen from $1 an acre to $190. At the same time the area was well on the way to becoming a major supplier of vegetables to Toronto and other southern Ontario cities and towns, as well as communities in the United States as far away as Philadelphia and the Jersey coast.

Today hundreds of thousands of motorists travel north and south on Highway 400 through this striking flatland, and the vast majority are unaware of the region's biologically rich wetland history. The fascinating cornucopia of life forms has been replaced by expansive fields of carrots and onions and celery – dusted with fertilizers, herbicides, and pesticides to produce the perfect-looking vegetables demanded by consumers who may not understand the environmental costs of the unblemished vegetables.

After years of studying and teaching biology and environmental studies, we know that the region is now ecologically sterile. It is, of course, a significant environmental loss. But for many years we too were unaware of the real story of what everybody called the "Holland Marsh."



Childhood recollections of driving through the "Marsh" still remain vivid. The coal-black soil made the young plants appear almost iridescent as they sprouted up in perfect green rows. The patterns formed by the neat rows and the patchwork of fields were like a huge quilt. For young city boys, the idea that the food on our table came from this mysterious land was intriguing, even mesmerizing.

Over the last four decades, we have watched the colour of the soil change ever so subtly. It is no longer as black as coal, nor as rich in texture. Some estimates suggest that the rich topsoil will be gone within one hundred years, and in some places only fifty years, lost through erosion, chemical oxidation, and attachment to vegetables when picked. What was built up over thousands of years may disappear in less than two hundred. The soil level in 1992 was falling away at the rate of approximately thirty centimetres (one foot) per decade – about 1.5 metres (five feet) since 1930.

Besides the loss of an important natural wetland, there are other environmental concerns in this intensively farmed region. Chemical insecticides, herbicides, and heavy metals such as copper and lead are applied to the soil and wash into the adjacent river and drainage canals. The residues accumulate in plants, fish, and other animals, especially those higher up the aquatic food chains, even though modern pesticides are less persistent than those used formerly. Attempts are being made to cut back, but concerns remain about the effects of bioconcentration and about related health issues for wildlife and humans.

The farmers, however, find themselves in a difficult position. While some consumers favour organic farming (and are willing to pay extra), the majority still demand "perfect" mass-produced vegetables. The Holland Marsh farmers, who must practise intensive agriculture on small plots, generally feel that they must have close to 99 percent planting success to survive. Biological controls, such as the release of predatory insects, are only about 30 percent successful. They don't do the job as well or as cost-effectively yet; therefore large amounts of chemicals remain in use.

Fertilizers are heavily used to make up for the loss of nutrients through erosion and the intensive market gardening. The increase in nitrogen and phosphorous in subsurface and runoff waters is a cause for concern in the Marsh, in the rivers downstream and canals, at an Area of Natural and Scientific Interest (ANSI), and at Cook's Bay of Lake Simcoe. The nutrient loading deprives the water of cleansing oxygen and leads to eutrophication and algal blooms – the same conditions that did so much to damage Lake Erie during the 1960s and 1970s.

The Holland Marsh experiment illustrates radical changes to a rich natural habitat and the chronic problems of intensive agricultural practices. While the Marsh supplies the consumer demands of the sprawling Greater Toronto Area just to the south, and beyond, more efforts should be directed at improved farming practices that address soil fertility and erosion and the environmental impact of chemical herbicides, pesticides, and fertilizers.

At the same time the loss of this biological treasure house is but one example of reckless and widespread drainage of wetlands in the Great Lakes–St Lawrence River area. The result has been the loss of much biological diversity and thousands of wildlife sanctuaries. During pioneer days almost all of this drainage was to provide agricultural land. Today the process continues as developers expand the boundaries of cities and towns and golf courses over wetlands. Many of these are in the resort and cottage areas of the Canadian Shield, such as the Kawarthas, Haliburton, and Muskoka – far from major centres of population.

The Holland Marsh experiment has lasted for seventy years and will probably go on at least another seventy before the rich topsoil is gone. It has provided vegetables to the urban population and a sometimes tenuous living to farmers. This has, arguably, been a far more useful enterprise than draining wetlands for golf courses. However, as we watch the land change colour from rich black to dark brown and contemplate the loss of wildlife and other environmental costs, we wonder if it was a wise undertaking.

Today, at a time when people are trying to save and restore wetlands in places like Toronto's Don River or the American areas of the basin, or elsewhere in the watershed, a proposal to drain a large and immeasurably rich wetland like the original Holland Marsh would meet resistance. And rightly so. We have learned some important things about nature since 1930 when the real marsh was drained – or have we? We like to think that we are at least beginning to learn.

# Wild Rice, Midges, and Cranes

## *The Importance of Wetlands for Creatures Great and Small*

*Scot Stewart and Bruce Litteljohn*

THEY ARE NOT MOUNTAIN SPIRES SLICING THE SKY, not deep canyons singing a river song. They are flat and mostly covered by water. Subtle rather than spectacular, they are places we have taken for granted. Many people view them as wastelands. We have drained them to create farmland, filled them to build towns and cities, and used them to bury our garbage, poisons, and unwanted materials. In most places, over half of them are gone, and in others 70 to 95 percent are lost. They are the Great Lakes and St Lawrence wetlands.

Flat, soggy, and often thick with aquatic vegetation, they have attracted few hikers, canoeists, and swimmers. Deceptively dull in some ways, they haven't drawn us to spend much time exploring them. They aren't a common destination for people wanting to dangle their toes in spring flowers or immerse themselves in autumn paint-pots of leaves. But all that is beginning to change.

Today we are starting to look at these marshes, lagoons, swamps, bogs, and fens with fresh eyes. And we are amazed, not only by what lives in the cattails, water lilies, bulrushes, and sedges above, but also in the submerged plants, muck, and rock below. According to the Nature Conservancy, almost half of the 131 plants and animals that are currently disappearing around the Great Lakes live only in these wetlands or are rarely found elsewhere. Names like white catspaw mussel, Illinois mud turtle, lakeside daisy, and Michigan monkey flower may catch our attention, but leave us wondering what they are and what value they have. Little noticed because they are nondescript or uncommon, they are like passing strangers – gone before we have a chance to meet them, let alone to know them.

East of the Great Lakes, between Montreal and Trois-Rivières, lies Lac Saint-Pierre. It contains 20 percent of the remaining wetlands along the St Lawrence River and is upstream from the salty tidal influence of the Atlantic. The wetlands there are home to most of the lake's seventy-nine different kinds of fish and are a vital stopping point for the clouds of migrating pintail ducks, snow geese, black ducks, and green-winged teals. As is the case with many wetlands, the fish that spawn in their shallows are important both ecologically and economically. Lac Saint-Pierre is one of the last strongholds of commercial freshwater fishing in Quebec. It also provides little kids and big kids with a chance to drown a worm or troll with something fancier: millions of dollars worth of fun and fish stories.



Other wetlands around the watershed are revealing equally important and fascinating stories. For example, we are starting to develop greater admiration and understanding of our "neotropical" migrants – smart birds, to us, the ones that beat it out of here in the draughty, snowflaky days of late September and October and head for the warmer climes of Brazil, Panama, Costa Rica, and Honduras. We have come to admire them because they do something we truly appreciate – eat bugs. They eat creatures we're not all that crazy about: leaf-eating caterpillars, flies, and mosquitoes.

Researchers followed some of these birds up the Lake Huron coastline in spring to see what they ate and where they hung out as they headed to their summer range. As they arrived on the north shore of Lake Huron, the birds were expected to move away from the cooler lake until the shore trees opened their leaves. Then it was thought they would move back to the shore to feed on the newly hatched insects munching on the young leaves. What surprised biologists was that insects called midges living in the warmer waters of the shallow rocky shoreline of the wetlands hatched and supplied the birds with all they needed to refuel. It showed how critically important it is to protect the birds' stop-over sites, as well as the summer and winter ranges. These overlooked places and times in animals' lives can prove critical to their survival.

Downstream, the water leaving Lake Huron flows into the St Clair River and then slows as it enters Lake St Clair. As it slows, the water loses part of the load of soil it is carrying. The result is a piecework of triangular, flat islands, including the spectacular Walpole Island.

Here, the First Nations people certainly know their ecological address and care deeply about it. Custodians of 6,900 ha (17,000 acres) of cattail, bulrush, and sedge marshland, as well as upland areas of tallgrass prairie and oak savanna, they stand guard over more than one hundred rare or endangered species of flora and wildlife. Their environmental programs and steadfast insistence on zero discharge of pollutants from the upstream industries of "chemical valley" are an example for us all.

Their work has been recognized on the world stage: beginning in 1993, the Friends of the United Nations began a planet-wide search for fifty communities that demonstrated an "outstanding collective approach to environmental issues and the social development of their inhabitants." Three hundred and sixty such communities were considered. After two years of study, fifty of them, from thirty-two nations, were selected. In 1995, at a glittering ceremony in New York City, the United Nations honoured the recipients of the "We the Peoples" awards. One of them was the Walpole Island First Nation – for its exemplary work in environmental research and sustainable development advocacy.

These descendants of the Ojibwa, Ottawa, and Potawatomi Nations are joined together in the Council of the Three Fires. The world-class wetlands of the St Clair River could not be in better hands, for what they do not already know about their ecological address through the traditional knowledge of the Elders, they find out through research, which includes working with non-native scientists and organizations.

They know that their inland delta wetlands form a rich nursery for fish and waterfowl but also provide for the water in other ways. The intertwining underwater plants and roots of shoreline plants act like huge pumps, drawing up nutrients and minerals from the river. At the same time, chemical pollutants and metals, especially lead and mercury – problem children of the upstream world of progress – are also pulled in and removed from the water stream, cleaning the flow to Lake Erie. There, the walleye fishery has seen a resurgence that underlines the need to lower contaminant levels in fish and wildlife. The impact and importance of wetlands is evident, especially as the flow into Lake Erie provides the drinking water for downstream towns and cities. Every natural filter helps.

At the west end of Lake Superior, in the heart of the continent, Ojibwa canoeists still paddle out into the shallow bays in August. As they ease into thick stands of tall, grassy plants they gently draw in the heavy tops and tap them carefully, releasing ripe wild rice into the bottoms of their canoes. The cold clear waters of Lake Superior seem so sterile that it is surprising they nourish such a crop – further evidence of the richness of marshes. After giving thanks for so prized a gift, the harvesters take the rice back to shore where it is turned into a perfect dish to accompany lake trout, whitefish, or duck. With care, the rice fields will continue to provide their bounty as long as new condominiums, ATVs, and personal water craft don't tear up the shoreline and the shallows, and the water remains clear and pure.

The sun has set over a series of lagoons and a river estuary at a spot where river and lake – this time, Michigan – meet. It is September, and mallards, coots, great egrets, black ducks, great blue herons, and perhaps a couple of hundred trumpeting sandhill cranes try to stock up for the great flight south. They have struggled to add every ounce of fat they possibly can before pushing on. Crimson crescents of light streak the water as the stragglers glide in. They have stretched the last remnants of daylight to eke out a final morsel or two before heading to safety – perhaps to open water – to roost for the night. A couple of final trumpets, a few quacks, then quiet. This tableau has played out for millennia across the wetlands, but it is hard to say how long such scenes will continue. Hard to say.

Facing page: Sandhill cranes. Photo: Scot Stewart

Acid precipitation is just one of the many types of air pollution that result from the combustion of coal, gas, and oil. When these fossil fuels are burned, they also release large quantities of carbon dioxide – one of the major greenhouse gases. The result is climate change. Global weather patterns are being altered: more severe storm events, warmer temperatures and drier conditions in some areas, which in turn can affect water levels in the Great Lakes–St Lawrence watershed.

Photo: Gregor G. Beck

# Covering the Acid Rain Story

## *The Insidious Drift of Air Pollution*

*Gregor Gilpin Beck*

*"Acid fallout can occur in any form – from rain, fog, and snow to dry deposition. The severity of the problem depends not only on the strength of the acid present but also on the sensitivity of the region receiving the pollution."*

DURING THE 1980S ACID RAIN was featured prominently by the media. Since then, however, decreasing coverage has been given to this environmental issue. Greenhouse emissions, ozone depletion, and rainforest clear-cutting – important issues in their own right – now grab more headlines. They also attract much of the limited funds available for environmental research. As a result, many people have assumed, incorrectly, that acid rain is no longer a serious problem.

The sources and effects of acid rain have been well documented. Sulphur dioxide, nitrogen oxides, and other pollutants released from automobiles and industry combine with moisture, oxygen, and light in the atmosphere to form an acidic solution. These acid clouds may travel either short or very long distances before precipitation falls back down. This same mechanism also explains the transport and deposition of many other forms of air pollution.

Acid fallout can occur in any form – from rain, fog, and snow to dry deposition. The severity of the problem depends not only on the strength of the acid present but also on the sensitivity of the region receiving the pollution.

Unfortunately, many areas most susceptible to damage are also downwind from major sources of pollutants. The Adirondacks in New York State and the Canadian Shield of northern Ontario and Quebec have granite bedrock with little or no capacity for buffering acid precipitation.

The catastrophic effects of acid rain were made plain decades ago with the documentation of innumerable dead and dying lakes. These waters appear as sparkling aquamarine jewels, but are virtually devoid of life. Clams, snails, crayfish, and aquatic insects – food for so many larger animals – do not survive, nor do the eggs and fry of fish. Although some adult fish can tolerate more acidic waters, there are limits. Ontario has about 250,000 lakes, half of which are sensitive to acid rain. Of these, 25,000 have suffered severe ecological damage, including the loss of many species of fish and other wildlife. The same is true for one-quarter of the lakes in the Adirondacks, and countless others in Quebec and abroad. Meanwhile on shore, salamanders, frogs, and toads lay eggs in pools of melted snow. The "acid shock" from an entire winter of acid deposition can lead to reproductive failure year after year. All are victims of a series of complex reactions involving water, soil, and numerous chemical elements such as calcium, aluminum, sulphur, and nitrogen. Sometimes, too, wildlife suffers from increased levels of toxic aluminum or methyl mercury which is released by the acidity.

The effects of acid rain on soils and vegetation are somewhat more subtle. Calcium, magnesium, and potassium are lost from the soil with acidic rainwater, reducing the availability of essential nutrients such as phosphorous and nitrogen. The result: loss of soil fertility which slows plant growth, increased susceptibility to drought and disease, and sometimes extensive die-backs of mature pine, maple, and other trees. While acid rain may not be entirely responsible for recent declines, there is little doubt that it is a major contributing factor in a complicated ecological equation.

Acid rain has been a major environmental issue since the beginning of the Industrial Revolution, when people first started burning large quantities of fossil fuels. With increasingly tall smokestacks, local conditions have often improved at the expense of other regions, other watersheds, and, not infrequently, other countries.

This clearly emphasizies the point that dilution is *not* a solution to pollution. Cleaner technologies and more stringent regulations for industries and automobile manufacturers have helped greatly to reduce the levels of acid-causing pollutants. But despite media coverage that waxes and wanes with fickle fashion, acid precipitation continues to fall. While the environmental damage it causes may no longer be getting worse in North America, the complexity of the issue ensures that the path to recovery will at best be very slow.

Photo: Frozenrope

# The Media Agenda

## *A Journalistic Perspective on Environmental Reporting*

*Michael Keating*

*"The media are the connecting rod between actions in the environment and reactions by the public. Since the 1960s the media have regularly confronted the world with ugly images created by a careless industrial and consumer society. Pictures of sewers, smokestacks, and leaking tankers defined the environment in the public mind."*

Most people get their environmental information from radio, television, and newspapers, so journalists have played a key role in shaping the way we see and think about our environment. The media are the connecting rod between actions in the environment and reactions by the public. Since the 1960s the media have regularly confronted the world with ugly images created by a careless industrial and consumer society. Pictures of sewers, smokestacks, and leaking tankers defined the environment in the public mind. Photographs and stories of dead fish, ugly sewer discharges, and garbage-laden lakes and rivers shocked the public and politicians into action to reduce the damage.

If the media help to set the public agenda on issues like environment, who sets the media agenda? Most news is coverage of announcements, public debates, and events staged by others. Journalists report what others say, so they rely on people to raise the issues. Environmental organizations, government, industry, and university professors are those most frequently quoted. When environmental groups said that the Great Lakes and St Lawrence were in serious trouble, this brought a considerable amount of coverage. When governments released similar messages, this consensus raised the public profile and credibility of the issue.

Conflict and drama, more than consensus, fuel the engine of the news machine. An environment story that involves people barricading a logging road, or toxic sludge gushing from a pipe, is seen as "news" because it involves conflict and change. Since they are constantly reacting to one breaking news story after another, few journalists have the time to do as much interpretive or follow-up reporting as many people would like.



The two other key factors that define "news" are time and space. Newspapers have the space to print only about 1 percent of the news stories they receive. In radio, a one-minute piece is a long story, and television loves speakers who can sum up a story in a ten-second news clip. Complex background or interpretive stories often get trimmed or deleted in the rush to cover new issues. Newspapers and newscasts are full of unrelated snapshots, the so-called slices of life. The result is a series of often disjointed stories rather than a coherent picture. Deadlines are the third powerful force shaping the news. Newspapers have at least one deadline a day, TV has several, and radio stations can have newscasts every hour. Sometimes the whole process of discovering a story and researching and writing it is finished in less than an hour. If the information is not available fast, it does not make the news.

Modern environmental journalism began in the 1960s, with a few stories such as DDT killing wildlife, and the "death" of Lake Erie. Environment reporting fledged in the 1970s with a growing list of pollution issues and took flight in the 1980s with the discovery of a seemingly endless litany of environmental crises, including dioxin in fish, holes in the ozone layer, deforestation, oil spills and nuclear accidents, climate change, the decline of fisheries, species extinction, and the human population explosion. With the Brundtland Report of 1987, journalists suddenly had to learn the lingo of economics and try to relate economic decision-making to environmental impacts.

How good a job have the media done? The major news media have given us a sense of the range of environmental problems, but often the coverage of issues focuses on the potential hazard without explaining the real risk, which only comes with exposure to the hazard. Too often, reporters simply get the science wrong, confusing, for example, the natural stratospheric ozone layer that protects us with the human-caused cloud of ozone that is caused by industry and cars, and harms our lungs down at ground level. Now that the media have laid out a great list of environmental issues for us to face, they need to provide the kind of in-depth analysis that would help us understand where to focus our attention. Without more analysis of the type that is applied to economics or political stories, we will continue to see the environment as a series of problems that are disconnected from one another and from the way we live and do business.

Stockton Island (Apostle Islands), Lake Superior. Photo: Scot Stewart

# The Great Lakes

# Reflections of a Paddle Pusher

## *Retracing Historic Canoe Routes to the Upper Great Lakes*

*Alec Ross*

*"I'll never forget the afternoon I emerged from the Voyageur Channel of the French River and entered the azure immensity of Lake Huron's Georgian Bay ... If I angled my bow in the right direction, I could gaze straight ahead into a horizon of nothing but water and sky. No land anywhere, not even in my peripheral vision. And that childhood sensation, a sort of thrilled vulnerability, hit me again: the exposure, the fantastic immensity of another Great Lake."*

I GREW UP A CITY KID IN TORONTO, and when I was in grade school the Great Lakes meant geography lessons and weekend ferry rides across Toronto Harbour to Centre Island. There I could play at an amusement park with ferris wheels, bumper cars, and a miniature train. If I wandered south, beyond the rides and the cotton-candy hoopla, I would soon reach the far side of the island where Lake Ontario – huge, blue, and dotted with screaming gulls – stretched towards the United States in the hazy distance. As a child I was amazed by the lake's size – so large that I couldn't see the opposite shore! I envied the people I could see on sailboats. My childhood memories of this Great Lake are fond, vivid, and lasting.

Many years later I was venturing on the Great Lakes myself, paddling a canoe along the north shores of Lake Huron and Lake Superior. My solo trip had begun in Montreal and taken me up the Ottawa River, where historic rapids were drowned by large hydroelectric dams built between the 1930s and 1960s. I portaged around six of them. Further upstream, I ascended the Mattawa River to North Bay, then crossed the height of land and followed the south shore of Lake Nipissing

Facing page: Photo: Frozenrope

to the French River. I was keenly aware that my route west had been used for hundreds of years by natives, and by later explorers such as Etienne Brulé, Samuel de Champlain, and others who charted those waterways for the countless missionaries, fur traders, and other travellers who paddled in their wake.

I'll never forget the afternoon I emerged from the Voyageur Channel of the French River and entered the azure immensity of Lake Huron's Georgian Bay, which, for good reason, is sometimes called the sixth Great Lake. If I angled my bow in the right direction, I could gaze straight ahead into a horizon of nothing but water and sky. No land anywhere, not even in my peripheral vision. And that childhood sensation, a sort of thrilled vulnerability, hit me again: the exposure, the fantastic immensity of another Great Lake.



Farther west again, I was paddling Superior, the greatest lake of all. Whirling through my mind were the horror stories I had heard about sudden squalls, mammoth waves, impenetrable fogs, and an often hostile, rocky coastline that in a storm could bash to pieces any boat that tried to land. The wreck of the *Edmund Fitzgerald* and other giant ships haunted me, and I felt even more tiny and exposed in my fifteen-foot canoe.

I had also been told of Lake Superior's spectacular scenery, but no words could have prepared me for the magnitude of its beauty and the feelings it evoked. Icy water so clear and seemingly clean that I could dip my cup over the gunwale and drink without fear of sickness; cobble beaches rising from the water in terraces, striking displays of the geological eras this lake has lived through; tangles of giant logs and branches piled up on shore, relics of mighty storms; cliffs and stones weathered smooth by eons of water lapping and crashing over them; colossal swells that lifted and dropped my canoe and made me feel as though I was riding a monstrous roller coaster. Everything here was evidence of timeless and boundless natural power. More than anything else, Lake Superior inspired profound respect.

Which is why, at Marathon, the sight of pulp mill effluent spewing a river of reeking yellow froth from a pipe into the lake hit me so viscerally. It was such a callous display of disrespect, an affront to the dignity of the lake. Superior was a living, breathing creature, and this effluent pipe was a cancer.

I thought of the Great Lakes further south, each of them fed by Superior. In recent decades they and the waterways that nourished them had been assaulted in similar fashion, with dire results all too familiar from newspaper and television reports: gulls and ducks with twisted, malformed beaks and eggs that never hatched; fish spotted with gross, fat tumours; dangerously high levels of toxic chemicals discovered in lake-bottom sediments; polluted beaches you couldn't swim at; oily rivers that caught fire.

These were extreme or perhaps historic cases, and some have been improved. But they underscore the depressing fact that when humans meddle with and defile the natural world, they are slow to learn from their mistakes. Those examples could still serve as harbingers of Lake Superior's future.

I crinkled my nostrils, trying to block out the sulphurous stench wafting from the pipe. It worked, sort of. But I could not suppress a desperate, involuntary prayer:

*Please God, not here.* Not in *this* Great Lake.

Osprey. Photo: Scot Stewart

# Poisons up the Food Chain

## *The Effects of Bioconcentration on Eagles, Ospreys, and Cormorants*

*Peter Ewins*

*"Eagles, ospreys, and cormorants, like other predatory species, are highly sensitive to chemical pollution and other forms of human disturbance. Because contaminants accumulate at each level in the food chain, these species are particularly vulnerable, and therefore act as warning signals for environmental health . . . Nonetheless, one must acknowledge the resilience of nature and the recovery of these birds since the restriction of the most persistent contaminants twenty-five to thirty years ago."*

THIRTY YEARS AGO, a summer boat journey along the shores of the Great Lakes and the St Lawrence River would have been very different from today. No bald eagles, ospreys, or double-crested cormorants. The reason? Breeding populations of these top predators had been all but eliminated by the far-reaching effects of toxic chemicals. Since the early 1970s, when use of many of the most harmful chemicals was restricted, breeding numbers of these three species have increased. Today, one can see again these magnificent predators along many shorelines – indicators of improving ecosystem health.

Eagles, ospreys, and cormorants accumulate toxic contaminants from the fish they eat. Their eggs provide a snapshot of the contaminants present in the body of the female at time of laying, and therefore usually reflect local environmental conditions. Many factors affect the size and health of populations. Over the past thirty to forty years, however, persistent organochlorine contaminants such as DDT and PCBs have had terribly detrimental effects. DDE has been particularly harmful. This by-product of DDT inhibits calcium deposition in developing eggs; the biochemical roadblock resulted in thin egg shells in the 1950s to 1970s, and eggs usually broke before hatching.

For bald eagle and osprey eggs, DDE concentrations of four parts per million or higher cause shells to be 15 percent thinner than normal, which significantly increases risk of breakage. Successive years of breeding failure meant that there were no new breeders to replace adults which died, so breeding populations declined and finally disappeared from most areas. Other contaminants such as dieldrin, PCBs, dioxins, and furans also adversely affected embryo development and adult survival.



During the 1980s concentrations of many toxic compounds in eagles and ospreys fell below critical levels, resulting in increasing populations and the recolonization of previously vacated territories. Since the 1970s, breeding populations of ospreys have increased dramatically in Lake Huron's St Mary's River and Georgian Bay, at rates of up to 15 percent per annum. There are now over 150 pairs of bald eagles nesting around the shorelines of the Great Lakes. In some heavily contaminated "hot spots," however, breeding success is still depressed, and chicks with bill and limb deformities have been observed, indicating high levels of PCBs.

Since European settlement, eagles and ospreys have been adversely affected by other factors. Industrial and residential development, particularly along shallow, sheltered shorelines, has reduced feeding and nesting habitat and dramatically increased human disturbance in many areas. The loss of large, dying trees eliminates nesting sites for both species, although artificial nest platforms are being used successfully.

Double-crested cormorants incubate eggs between their feet where thin shells are easily crushed, so DDE poisoning was a particularly severe problem. Despite continual persecution at breeding colonies, there were nine hundred breeding pairs on the Great Lakes in 1950. By the early 1970s, however, only one hundred pairs remained. During this period, eggshell thickness decreased by up to 30 percent. The population has now increased to over 40,000 pairs. The spectacular increase in cormorant numbers is due to many factors, including reduced contaminant levels, increased legal protection, and the increased availability of forage fish attributable to the decimation of predator fish populations by over-fishing and the parasitic sea lamprey.

Like bald eagle chicks, cormorant young can also exhibit malformations, perhaps induced by PCBs, dioxins, and furans. The average incidence of cormorant cross-bill deformities in the Great Lakes has decreased to 0.02 per 10,000 chicks. However, in Green Bay, a highly contaminated region of Lake Michigan, deformed chicks are still common: fifty-two per 10,000 chicks in 1979–87, increasing to seventy-three per 10,000 in 1989–95.

Eagles, ospreys, and cormorants, like other predatory species, are highly sensitive to chemical pollution and other forms of human disturbance. Because contaminants accumulate at each level in the food chain, these species are particularly vulnerable, and therefore act as warning signals for environmental health. Over the past fifty years, toxic chemicals have had a far greater impact on the breeding success of these birds than any other single factor. Nonetheless, one must acknowledge the resilience of nature and the recovery of these birds since the restriction of the most persistent contaminants twenty-five to thirty years ago.

# Rescuing Native Fish

## *Restoration of Lake Trout Populations in the Great Lakes*

*Thomas A. Edsall*

*"Restoration of the lake trout is an important activity in its own right. However, its importance can be appreciated more fully if it is remembered that the loss of the lake trout fishery in the 1940s and 1950s sparked not only this specific recovery project but also a concerted effort to reduce pollution throughout the Great Lakes-St Lawrence drainage basin."*

THE LAKE TROUT is the largest predatory fish native to the Great Lakes, St Lawrence, and many other cold northern waters in the United States and Canada. It is long lived and slow growing, weighing seven kilograms (fifteen pounds) when it reaches ten to fifteen years, although individuals many times that weight have been recorded. Because of its large size, historic abundance, and predatory habits, the lake trout formerly exerted a major influence on the structure and stability of cold-water fish communities in the Great Lakes.

The size and rich flesh of lake trout made it a popular commercial fish, particularly in Lakes Superior, Huron, and Michigan, where about seven million kilograms (fifteen million pounds) were landed annually from the late 1800s through to the mid twentieth century. Populations in the Great Lakes collapsed in the 1920s to 1950s due to prolonged, intensive commercial fishing and to predation by the parasitic sea lamprey. Stricter regulation of the commercial fishery together with a sea lamprey control program begun in the 1950s spared remnant populations from extinction in Lake

Facing page: Photo: Frozenrope

Superior and parts of Georgian Bay. However, native lake trout populations elsewhere in the Great Lakes were lost.

Since the late 1950s, hundreds of millions of juvenile lake trout have been stocked in all five lakes in an attempt to re-establish abundant, naturally reproducing populations throughout previously occupied habitats. The recovery of self-sustaining populations is nearly complete in Lake Superior, but less success has been realized in the other four Great Lakes.

In northern Lake Huron, stocked lake trout established breeding populations in the 1980s, but fishing regulations did not adequately protect this spawning population, and it has been lost. In 1994 large numbers of mature, stocked trout produced significant numbers of young for the first time on Six Fathom Bank–Yankee Reef complex in south-central Lake Huron. This 520 square kilometre area (two hundred square miles) was provisionally established as a lake trout sanctuary in the early 1980s and accorded full sanctuary status in 1997. Similar sanctuaries established in Lake Michigan now support strong populations of breeding-age fish. Successful natural reproduction should be evident soon in Lake Michigan if present fishing regulations and sea lamprey controls are maintained.

The production of newly hatched fry by stocked lake trout is now widespread in Lake Ontario, and naturally spawned fish are now present in the population. However, predation by alewife, a species which may have been accidentally introduced into the Great Lakes in the 1800s, is believed to be limiting fry survival.

Restoration of the lake trout is an important activity in its own right. However, its importance can be appreciated more fully if it is remembered that the loss of the lake trout fishery in the 1940s and 1950s sparked not only this specific recovery project but also a concerted effort to reduce pollution throughout the Great Lakes–St Lawrence drainage basin. This latter, remarkably successful effort has resulted in major improvements to water quality and ecosystem health, which has contributed significantly to the development of a multi-species fishery resource, with an estimated annual value of more than four billion dollars.

Native fish species face many challenges. They must cope with pollution and with the pressures of sport and commercial fishing. An additional threat to native flora and fauna comes from the tremendous abundance of non-native species. Some have been introduced intentionally, including Pacific salmon and rainbow trout; others have arrived accidentally, such as the exotic ruffe and goby fishes, and, of course, the now infamous zebra mussel. These diverse factors make the job of understanding and managing aquatic ecosystems increasingly difficult.

The Lake Erie shoreline Photo: Gregor G. Beck

# Making the Lakes Great

## *The Role of Remedial Action Plans*

*Gail Krantzberg*

*"Remedial Action Plans have rehabilitated hundreds of kilometres of riparian [i.e., riverside] vegetation and thousands of hectares of wetlands. Sediment quality has been vastly improved in many locations because of pollution control and clean-ups. More fish are edible in more places, and swimming is again possible in parts of our urban centres for the first time in decades."*

In 1909 reasonable people recognized that water does not abide by politics. The Boundary Waters Treaty signed by Canada and United States was the first attempt to protect the shared waters. More recently, the two nations developed the Great Lakes Water Quality Agreement and made a commitment to restore and maintain the biological, chemical, and physical integrity of the waters of the Great Lakes Basin Ecosystem. The 1987 Agreement specifically called for the development of Remedial Action Plans (RAPs) at forty-two Areas of Concern where deterioration was particularly pronounced. RAPs were to embody a systematic, comprehensive ecosystem approach to restoring and protecting the Areas of Concern. Further, the federal governments, in cooperation with state and provincial governments, were to ensure that the public was consulted throughout the development and implementation of the plans.

I have worked with the Canadian RAPs in an attempt to quantify the step-wise progress made in restoring environment quality, and to reach an understanding of the challenges ahead. Ten years after their inception, RAPs have rehabilitated hundreds of kilometres of riparian vegetation and thousands of hectares of wetlands. Sediment quality has been vastly

Facing page: Photo: Frozenrope

improved in many locations because of pollution control and clean-ups. More fish are edible in more places, and swimming is again possible in parts of our urban centres for the first time in decades. Conservation and protection and watershed and land-use planning are becoming central to sustaining the gains achieved. Some of the RAPs are about to move into a monitoring mode, tracking the natural recovery anticipated as a result of their efforts. Tens of thousands of volunteers are giving their energy to revitalize and monitor their domain. Scores of funding partners have collaborated. Research is being advanced across the watershed on the insidious nature of endocrine hormone disruptors and other chemicals. Technologies are emerging for efficient management of stormwater and wastewater, sediment remediation, and greener industry. These are exciting times for the lakes.

These are also particularly challenging times for the lakes. Many institutions direct virtually all their efforts and funds remaining from shrinking budgets to reducing loadings of chemicals to the lakes. The integrity of the Great Lakes ecosystem, however, is sharply challenged by the massive loss of habitat, reckless land-use decisions, and invasion of exotic species. The research efforts of Canada and the United States have placed an emphasis on examining chemical problems, reflecting the tendency for agencies to look towards technology for immediate solutions. Far less diligence has been directed at reaching new standards of conduct to redress careless land use, habitat destruction, resource extraction, and biological imbalance.

Other stressors on the lakes such as global warming, development, and declining species biodiversity cannot be solved at the local level. Here the challenge is to launch inter-organizational programs to provide the fabric within which site-specific actions can be cultivated. Government agencies and university consortiums, together with international organizations, should be promoting and adopting a broader research dimension that will result in the cooperative pursuits necessary to grasp ecological integrity.

We are only now beginning to evaluate the degree to which environmental quality has been and indeed can be restored in the complex of social, economic, and political variables. A clear articulation of the economic and health consequences of investing in Great Lakes restoration and protection (or not) is urgently needed. We must be able to measure and extol the benefits of improving the lakes for our pockets, our children, and our souls.

Beyond the need for research and technology is the need to sustain the commitment of those who participate in the process – both the agencies and the public. The passion and dedication of those involved in the RAP process are rarely articulated or presented in the printed word. The solidarity that has emerged in many of the Areas of Concern sees former adversaries become allies united by the vision of a shared inspiration. Recently, several agencies have come close to eliminating all staff dedicated to restoring the Great Lakes, yet some of these experts choose to continue their meritorious work on their own time or by other innovative arrangements. In the face of being declared redundant, they rise to the challenge and continue to look forward rather than abandoning the campaign. This level of commitment and dedication speaks to the power of the RAP process, the unification of persons charged with the aim of enhancing and protecting the magnificence that is the Great Lakes. While local associations are forming to respond to the funding squeeze, governments must honour their responsibilities as full members of the Great Lakes community.

We must foster new ways to ensure that we do not lose the passionate and driven individuals who help to weave the web. This includes the community groups that have brought together local leaders to restore their corner of the Great Lakes. Creative, innovative partnerships and institutional arrangements are needed to advance the successes being measured throughout the basin. These take time. The time affords agencies the ability to protect the investment of hundreds of millions of dollars spent to date in controlling eutrophication, restoring riparian vegetation, rehabilitating coastal wetlands, and cleaning up contaminated sediment. Through the RAP process, we have been raising public awareness of individuals' responsibilities, forging partnerships, engaging tens of thousands of volunteers, uniting municipalities, and making a difference. In the end, the majesty of the lakes will be the virtue that sustains the passion of all of us who live and grow in this wondrous place.

# Developing Sustainability

## *Conservation Action in the Lake Superior Basin*

*Gail Jackson and Bob Brander*

*"A number of citizen-based planning initiatives have also emerged in the Lake Superior watershed. The one thing they have in common is concern – for the future of their community, their traditions, their children and their grandchildren ... When local people are directly involved in a process which empowers them to make decisions affecting their future, the initiative has the greatest likelihood of success."*

LAKE SUPERIOR has been called the crown jewel of the Great Lakes. This comes as no surprise to those who have paddled the coast or camped upon the many beaches scattered along more than 4000 km (2,700 miles) of shoreline. There are areas here that appear to be in a primordial state, seemingly untouched by humans. Lake Superior inspires a sense of awe and wilderness, and commands respect.

Human habitation and development have not been imposed upon Lake Superior as much as on the other Great Lakes. National, provincial, and state parks acknowledge the natural beauty and splendour of the Superior landscape. More than one and a half million hectares within the watershed are protected by nine parks alone: clockwise from the most northerly reaches of the basin, Wabakimi Provincial Park, Pukaskwa National Park, Michipicoten Island Provincial Park, Lake Superior Provincial Park, Pictured Rocks National Lakeshore, Porcupine Mountains State Park, Apostle Islands National Lakeshore, Isle Royale National Park, and Sleeping Giant Provincial Park.

Other designations contribute as well to this impressive

Facing page: North shore, Lake Superior. Photo: Frozenrope

array of protected areas. But not all natural regions are represented, and there has been a conspicuous absence of marine protected areas. Furthermore, protected lands isolated from one another are not enough to retain the extraordinary character and integrity of the lake and its watershed. Nor will they satisfy the long-term needs of the people living and working in the basin. A conscious effort must be made to manage the Superior ecosystem as a whole.

Managing large lake ecosystems such as Lake Superior is a formidable task. Policies, procedures, and the actions of two federal governments, three state governments, and one provincial government need to be coordinated. Cooperation and support of regional industries, communities, stakeholder groups, and residents must be elicited, and the special interests of Aboriginal peoples must be taken into consideration. In spite of these complexities, there is one key advantage to managing an ecosystem the size of Superior: it can accommodate a balance between conservation measures and resource use. This philosophical approach has been called *sustainable development*. By reversing the order of this popular two-word phrase to *developing sustainability,* one can eliminate the apparent contradiction.

In the 1970s and 1980s, concern was raised over the future of Lake Superior. Eight areas were identified as being particularly at risk: Thunder Bay, Nipigon Bay, Terrace Bay and Jackfish Bay on the north shore, St Louis River/Duluth Harbour, Deer and Torch Lakes on the Keweenaw Peninsula, and St Mary's River which connects Lake Superior and Lake Huron. Remedial action plans were launched to restore these Areas of Concern. In 1989 the International Joint Commission challenged governments to designate Lake Superior as a demonstration of zero discharge of persistent toxic substances. The governments of Canada, the United States, Ontario, Michigan, Wisconsin, and Minnesota responded to this challenge by announcing a "Binational Program to Restore and Protect the Lake Superior Basin" in 1991. Superior was chosen on the premise that if the lake least affected by industrialized society could not be protected, there would be no hope to restore and protect the other four Great Lakes.

The program hopes to deliver *zero discharge* of persistent, bioaccumulative toxic substances through innovative pollution prevention strategies, enhanced control and regulatory measures, and special designations, as well as a program of progressive remedial actions and long term watershed-wide management planning. Communities, industry, academic institutions, environmental organizations, and special-interest groups are actively participating and contributing to the success of the Lake Superior Binational Program. This program recognizes that healthy communities depend primarily on the natural resources of the surrounding area. An approach to maintain the integrity of the ecosystem is evolving – one that sustains the environment, the economy, and communities.



Canada is a nation with a rich maritime heritage that spans three oceans and four Great Lakes. It is Parks Canada's goal to establish a network of national marine conservation areas (NMCA) representing each of Canada's twenty-nine marine regions. The program is in its infancy with only three NMCAs established to date. These conservation areas are intended to represent the abiotic, biotic, and cultural attributes of the marine region and are managed to achieve ecological sustainability. They embrace a number of management concepts including protection, sustained use of natural resources, research, and education. As such, NMCAs have a similar philosophical approach to the Lake Superior Binational Program but are applied at a regional scale.

With attention already focused on Lake Superior through the Lake Superior Binational Program, Parks Canada undertook a regional assessment to identify NMCA candidate sites. The waters associated with the island archipelago near Black Bay Peninsula and easterly to the Slate Islands have the greatest potential to represent the Lake Superior marine region. The federal and provincial governments launched a feasibility study to determine if there is support for a NMCA proposal – unless the hearts and minds of north shore residents support the proposal, it is not likely to succeed. The process is now considering possible boundaries for this new protected area.

Pukaskwa National Park represents Canada's central boreal uplands forest natural region. The park ecosystem is intrinsically linked to the surrounding hinterland. In recognition of this, adjacent land owners, managers, and stakeholders have been invited to participate in a regional planning exercise. Pukaskwa has hosted two ecosystem management workshops to provide a forum for potential partners to discuss common goals and collaborate on joint initiatives. Efforts have been focused on planning for sustainability, regional communications, and ecosystem conservation. Workshop participants recognize that partnerships have the potential to enhance expertise, reduce costs, and facilitate a more objective approach.

Through the efforts at Pukaskwa National Park, a regional steering committee has been established and cooperative projects are beginning to emerge. One of the largest collaborative research projects is a predator-prey study involving the Ontario Ministry of Natural Resources, the Forest Industry, and Lakehead University. Habitat use and the interactions of moose, caribou, and wolves are being studied in the park and surrounding area. Other initiatives include water-quality monitoring on the east and west Pukaskwa Rivers and the White River, strengthened communications with the Hemlo mining community, and development of a newsletter. Linkages to adjacent industries, surrounding communities, and research institutions will ensure Pukaskwa National Park does not become an island isolated from the economic and social realities of its neighbours. [With the signature of the Ontario Forest Accord in 1999, a large new protected area was created between Pukaskwa and Lake Superior Parks, helping to ensure that they will not become isolated "ecological islands." In total, over 500,000 hectares of new conservation reserves and parks were created on the Ontario side of the Lake Superior watershed in 1999. The Ontario Forest Accord, which stemmed from the Ontario government's controversial Lands for Life process, also created significant new protected areas along the eastern and northern shores of Georgian Bay and Lake Superior, the "Great Lakes Heritage Coast." *Eds.*]

A number of citizen-based planning initiatives have also emerged in the Lake Superior watershed. The one thing they have in common is concern – for the future of their community, their traditions, their children and their grandchildren. Simply put, they share an inherent desire to control their own destiny. When local people are directly involved in a process which empowers them to make decisions affecting their future, the initiative has the greatest likelihood of success.

One such example is the village of Rossport, located centrally on Superior's rugged north shore. The area is rich in scenic beauty, natural resources, and recreational opportunity. Community action was catalysed in 1988 when area residents reviewed a tourism strategy which proposed offshore tourism development on the pristine island archipelago. They raised strong objections to the proposal, primarily because of its potential to increase pressure on fisheries resources, degrade fragile island ecosystems, and contribute little to the economy of the community. A comprehensive long-term plan for the Rossport Island Area was urgently required.



With support from the provincial government, a pro-active Vision 2000 workshop was held in 1990. The Rossport Area Conservation and Development Group formed to initiate studies for a marina and mainland-based tourism opportunities. A second group, the Rossport Island Management Board, also emerged to create a long-term plan for managing the islands. Local citizens invested considerable time and effort on the initiative and drafted a management plan in 1994. The process provided tremendous insight into issues and concerns of local citizens and offered guidance to the provincial government for managing Crown islands. It is hoped that community action will result in a secure, sustainable, and healthy future for the residents of Rossport and the island archipelago to which they are inherently connected.

On the opposite shore, the Keweenaw Peninsula represents a spectacular section of Michigan's Superior coast. In 1988 a group of developers proposed a $1.2 billion pulp mill for the area. Concerned with the proposal, some local citizens protested, and developers ultimately withdrew their proposal. With this initial success, the Friends of the Land of Keweenaw formed to have a greater influence over activities on the Keweenaw Peninsula and to create a sustainable development plan.

Although well intended, the process was not all-inclusive: economic development interests were not part of the discussions which preceded the plan. As a result, when the plan was announced, potential developers refrained from discussions on how to implement it. Without the willing involvement and full participation of the developers, the plan could not be activated. "Sustainability" advocates and economic developers began to engage in more productive discussions in 1993. The future of the

Keweenaw Peninsula is now brighter as all stakeholders strive to achieve a balance between conservation objectives and resource use.

Chequamegon Bay is on Wisconsin's Lake Superior coast, about 160 km (one hundred miles) southwest of the Keweenaw Peninsula. The area is poised for renewed economic development, particularly through tourism and logging-related industries. Early in 1993, local citizens initiated discussions on sustainable development in the Chequamegon Bay region. They examined grassroots sustainability initiatives in other Great Lakes regions, including the Keweenaw Peninsula. Building upon successes and learning from failures, they moved cautiously but steadily forward. Economic developers and regional elected officials were brought into the discussions. By late 1994 the discussion group became the Alliance for Sustainability, with the enthusiastic endorsement of the region's three mayors and chairpersons of the two Chippewa tribal communities.



The Alliance functions through several committees, including a committee on economic development with members from local business and industry. Other committees, such as ecosystem management, review development proposals in terms of sustainability. Chequamegon Bay citizens are beginning to demonstrate that "sustainability" and "development" need not be conflicting concepts.

Sustainability, by definition, must endure. A symbiotic or mutually beneficial relationship must be developed between the human community and its habitat. The first step for the human community is to recognize the cultural and economic connectiveness of this bioregion. The second step is to achieve social, economic, and environmental harmony. It is no longer a matter "if" this can be achieved, but "when" it can be achieved. Selfish and short-term ambitions must be set aside in favour of long-term sustainability. It is a big step for the new millennium – a step of global proportions – but one that must endure the test of time.

Great Lakes Heritage Coast (Georgian Bay) Photo: Gregor G. Beck

# From Apathy to Action

## *Environmental Issues in American Heartland Cities*

*Rae Tyson*

> *"In any of the lakes the most obvious change involves waterfront real estate. Cities have finally realized that lakefront and riverfront properties and beaches are assets to be exploited. Indeed, in cities like Buffalo, Milwaukee, Cleveland, and Detroit, the waterfront has played a pivotal role in the revitalization of the inner city."*

To grow up along the Great Lakes in the 1960s was to witness an ecosystem near death. The decade was a blur of closed beaches, burning rivers, foul water, and dead fish. Obituaries for Lake Erie appeared regularly in magazines and newspapers nation-wide.

To cover the Great Lakes as a reporter in the 1970s and 1980s made the situation seem worse. The biennial water-quality report of the International Joint Commission (IJC) pin-pointed so many areas of concern that the U.S. coastal maps – from the western tip of Lake Superior in Minnesota to the eastern end of Lake Ontario in New York – were nearly oblit-erated by black dots.

The poisonous insecticide DDT was found in Lake Superior and other lakes, the result of agricultural runoff – and atmos-pheric deposition. Polychlorinated biphenyls (PCBs), a family of poisonous compounds once widely used in electrical equip-ment, permeated the entire lake system. The highly toxic by-product commonly called dioxin – from paper mills and chemical dumps – tainted the flesh of fish and waterfowl, par-ticularly in Lake Ontario. Life-choking nutrients from farm

Facing page: Photo: Phillip Norton

runoff and human sewage nearly killed Lake Erie, the shallowest and most vulnerable of the lakes.

And, as William Ashworth accurately pointed out in his book *The Late Great Lakes*, the whole situation was exacerbated by tremendous apathy among government agencies and citizens alike.

The apathy involved more than dirty water. In city after city, what should have been the most valued urban land was treated instead with disdain. In Indiana, Ohio, and Illinois, the lakefront was dominated by steel mills, petroleum refineries, and other factories and warehouses. In New York State, the Niagara River linking Lakes Erie and Ontario was in essence used as an industrial sewer when petrochemical plants built leaky dumps at water's edge.

The best-known of those dumps was, of course, the notorious Love Canal site in Niagara Falls. Used by the Hooker Chemical Co. (now Occidental) until the mid 1950s, the dump began leaking into nearby homes, forcing a widespread evacuation and cleaning beginning in 1978. Love Canal, itself a major source of dioxin and other persistent contaminants, was only one of four Hooker sites that contributed to the tremendous pollution problems in the Niagara River and downstream in Lake Ontario. Final cleanup at all four sites could extend well into this century.

So much for history. The greater question is: does apathy still prevail? From a more distant vantage point in Washington, D.C., the answer would seem to be "no." Others much closer to the situation say "maybe." "One of the hardest things to overcome is that apathetic attitude," says Indianapolis lawyer Gordon Durnil, U.S. chair of the IJC until 1994.

In any of the lakes the most obvious change involves waterfront real estate. Cities have finally realized that lakefront and riverfront properties and beaches are assets to be exploited. Indeed, in cities like Buffalo, Milwaukee, Cleveland, and Detroit, the waterfront has played a pivotal role in the revitalization of the inner city.

Perhaps the most dramatic is the renaissance in Cleveland where a new waterfront baseball stadium and rock and roll museum have helped stimulate dramatic improvement in the downtown area. And the Cuyahoga River is still hot – though no longer from burning pollutants: the confluence of the river with Lake Erie, lined with restaurants, bars, and upscale shops, has become the magnet for the city's young professional workforce. "The transformation has been amazing," says Jim Fryan, owner of *Goodtime III*, a Cuyahoga/Lake Erie tour boat.

Detroit's efforts at a similar transformation along the Detroit River have been slower to materialize, but Buffalo has used its waterfront as the basis for a major downtown revitalization effort. Buffalo's waterfront now has restaurants, hotels, and a sizeable residential development.

But Durnil says Chicago is a good illustration of the dichotomy that exists throughout the Great Lakes on the American side. Though the downtown waterfront with its miles of groomed beaches is well maintained and accessible, the lakefront to the east towards Whiting, Indiana, is still dominated by refineries and other industrial sites. Finding waterfront success stories, says Durnil, "depends on where you go."

Water quality progress also has been uneven for a variety of complex reasons that include public apathy, lack of resources, unclear jurisdiction and, in a number of cases, realization that lakes cleanup is an extremely complex process. "The problem is much, much, more intense than anyone ever realized," says Jack Weinberg of Greenpeace in Chicago.

Green Bay, Wisconsin, along the western shore of Lake Michigan, is a good case study. For years a variety of municipal and industrial sources – including leaky landfills and paper-mill discharges – contributed to the degradation of the bay and, ultimately, Lake Michigan. Two decades of regulatory activity by Wisconsin and neighbouring Michigan have now resulted in upgraded municipal sewage treatment, tougher discharge standards for industry, bans on phosphate detergents, and cleanup of several leaking disposal sites. Despite those efforts, the area is still one of concern because, among other things, farm runoff into bay tributaries continues to be a major source of pollution. Though the problem is by now well documented, some farmers have resisted new regulations, and some local governments have been slow to impose them. Addressing the problem has been complicated because a number of state and local government jurisdictions must cooperate to find a solution. Consequently, hazardous pollutants still flow into the bay – and consumption advisories are still in effect for fish caught throughout the western end of the lake.

"We've come a long way, but we still can't eat the fish," says Ken DePrey, a sportsman in Dyckesville, Wisconsin.

Green Bay is also affected by another typical problem: sediment, tainted by PCBs and other pollutants, remains at the bottom of the bay, then to be stirred up whenever there is a turbulent storm. Estimates for removing the sediment exceed $500 million.



And it was farm runoff that may have been responsible for a public-health disaster in 1993, when many dozens died and over 400,000 sickened because of drinking water tainted with the microbe *Cryptosporidium* in Milwaukee.

Unfortunately, the problems in Green Bay and elsewhere in Wisconsin are repeated throughout the system. And the costs to address these historical problems are staggering, particularly in an era of cost-cutting in Washington, and in statehouses throughout the Great Lakes.

In Milwaukee the problem is similar, though exacerbated tremendously by encroaching development and urban land-use practices that do little to curb runoff. In Buffalo, Detroit, Chicago, and Cleveland, the scenario is repeated. "People ought to be able to fish free from fear and swim free from fear," says Shirley Tomasello of the Lake Erie Alliance. "Unfortunately, they can't."

Great Lakes cleanup has also been hampered by a string of new discoveries. The latest is a growing debate over chlorine, a substance widely used by industries. Environmentalists say past and present uses of chlorine are responsible for widespread problems ranging from deformed bald eagles in Lake Erie to elevated cancer rates among fish-eating First Nations throughout the Great Lakes watershed. Industry is resisting calls by the IJC and environmentalists to ban most chlorine uses.

As the cleanup of the Great Lakes moves forward in fits and starts, Durnil sees a marked contrast between the United States and Canada. In Canada, much of the nation's population lives in close proximity to the lakes. In the U.S., the opposite is true, a fact that has continuously handicapped efforts to make the Great Lakes clean-up a national priority. "The political power [of Canadians] lives there; the political power of our nation moved away from there," Durnil says.

Bruce Johnson of the Lake Michigan Federation agrees. "If you can't generate the political will, [cleaning up] the Great Lakes is going to be a tough sell."

# The Crush of Megalopolis

## *Crowding Nature in Ontario's Golden Horseshoe*

*Brad Cundiff*

*"The Golden Horseshoe has become the densest urban conglomeration in a country where the vast majority of people live in cities. Dense, mature hardwood forests have been replaced by pavement — practically none of the area's original forest cover remains. More than 80 percent of the area's wetlands have been destroyed, and rivers and streams have been denuded of shoreline vegetation, channelled, and even buried underground in pipes."*

CRADLING THE WESTERN END OF LAKE ONTARIO, the Golden Horseshoe represents Canada's industrial and commercial heartland. It stretches along two hundred kilometres (125 miles) of shoreline and includes a large, rapidly urbanizing surrounding area. The name "Golden Horseshoe" is itself full of the 1950s-style optimism of industrial progress, recalling that decade when the connections between Toronto, Hamilton, and Niagara were pulled tight by the building of the country's first expressways.

The new arteries brought with them a development style and attitude that eliminated the once-clean edges of the Horseshoe's urban areas: suburbs, strip malls and industrial "parks" spilled out along the new road links. These developments in turn gobbled up the fertile land and rich woods that had led European immigrants to concentrate their settlements around the natural harbours of Toronto and Hamilton in the first place.

The Golden Horseshoe has become the densest urban conglomeration in a country where the vast majority of people live in cities, and it is home to about 20 percent of Canada's population. This great density is most evident from Oshawa

Facing page: Winter skiing. Photo: Gregor G. Beck

and Toronto westward to beyond Hamilton. Glimpses of rural life may remain here and there, but it is not really possible now to leave the city behind in travels through the corridor.

The impact of this large urban area on the lake it adjoins is both straightforward and complex. At the relatively straightforward level are the direct and indirect toxic discharges from thousands of industries, both large and small, into the lake. These combine with pollutants from a multitude of other sources – from landfill leachate to household wastes to oil dumped down storm sewers – all forming a nasty cumulative brew just offshore.



The area has also undergone profound physical changes. Dense, mature hardwood forests have been replaced by pavement – practically none of the area's original forest cover remains. More than 80 percent of the area's wetlands have been destroyed, and rivers and streams have been denuded of shoreline vegetation, channelled, and even buried underground in pipes. Along the edges of Lake Ontario itself only a fraction of the generally soft, sandy-clay shoreline remains. Property owners, anxious to protect their holdings, have armoured the shoreline for most of its length with cement break-walls and rock.

During the economic boom of the 1980s and continuing today, agricultural lands have taken a beating. Between 1981 and 1986, a peak period for new suburban development, one hundred square kilometres (thirty-eight square miles) of prime agricultural land in the Toronto region alone were converted to mostly low-density development.

At the southeastern end of the Golden Horseshoe, the Niagara River spills into the lake, the major source of the hydroelectric power that originally drove the growth of industry in the region. Notorious since Love Canal, the Niagara is also the source of 80 percent of the water that enters Lake Ontario. Those waters take six years to travel through the lake and are vital as a source of drinking water.

Further west, Hamilton Harbour is hoping to become a remedial success story. Industrial discharges from a large steel-making industry have been reduced, and the removal or capping of often-toxic sediments in the harbour is well underway. Large new holding tanks are helping Hamilton to deal with its sewage; but the method is controversial since some environmentalists feel this is only a "Band-Aid" solution that fails to address the root of the problem.

The harbour still has a long way to go, however, especially in terms of habitat re-creation. What was once one of the most productive natural bays on Lake Ontario now offers only pockets of natural refuge, and these contain a fraction of the complexity of the original "paradise" which settlers discovered in Hamilton.



Sprawling from its core to the north, west, and east, Toronto is Canada's most populous metropolitan area. Like Hamilton, Toronto has problems dealing adequately with sewage and stormwater runoff. In fact, Toronto's sewage system has put the city on the map – the International Joint Commission's map of Great Lakes "hot spots." Toronto's outdated sewage system, parts of it dating back to the 1800s, has proven incapable of dealing with both the volume and content of the waste it now receives. Ironically, the potential burden on sewage systems has in some cases limited further developments and slowed the path of the bulldozer.

Toronto marked the beginning of the era of industrial progress by turning the huge river-mouth marshes of the Humber, and especially the Don, into fetid cesspools, convenient for the dumping of all sorts of waste. The city's solution to this unpleasant problem was to fill in (or "reclaim") the Ashbridge's Bay marsh at the mouth of the Don in the early twentieth century and to start construction on a sewage collection and treatment system. Unfortunately, the complex problems that bedevil Toronto today date right back to those early engineering efforts.

The problem, and Toronto is far from alone in facing it, is that as the city grew, a system of combined stormwater and sanitary sewers (for sewage and household waste-water) was built beneath it. Single sewers carried both stormwater and "sanitary" waste-water to treatment plants, but in times of heavy rainfall or excessive water use, the system can be overloaded. This "combined sewer overflow" (or CSO) is then diverted away from sewage facilities and enters streams and the lake untreated.

While combined sewer overflows remain a major problem in the older parts of Toronto, most newer sections of the Golden Horseshoe have separate storm and sanitary sewers. In these cases, stormwater usually enters the watershed untreated, while sanitary sewers take sewage to treatment facilities. Improper hook-ups and antiquated systems, however, cause chronic and uncharted problems: stormwater enters sanitary sewers, contributing to overloading, and no doubt some sewage still enters storm sewers directly. To further complicate matters, the separate sewers are cross-connected in many places to allow the sanitary sewer to overflow into the stormwater sewer, usually during heavy rainfalls. Again, the end result is that raw sewage, industrial waste, and household waste-water mix with stormwater and end up entering the rivers – and ultimately Lake Ontario – untreated.

On a set of blueprints, the solution to these problems seems straightforward: separate all sewer systems, increase the capacity of the sanitary system, and disconnect the cross-connections. And that, in fact, is what municipalities throughout the Golden Horseshoe, including Toronto, have spent the last decade or more doing. But as former Toronto Remedial Action Plan Coordinator Ed Sado notes, that is just the beginning. First of all, he points out, "you will never find all the cross-connections." Many older buildings and homes have cross-connected drains, he explains, and only if and when such buildings are redeveloped will the situation be discovered.

Even if a perfect split could be achieved, Sado adds, the quality of stormwater flowing directly into the lake, and indirectly through feeder streams and rivers, still leaves a lot to be desired. This, after all, is not rainwater or snow melt that has slowly filtered through the dense web of a forest floor – it is the flash runoff from oil-soaked asphalt and gritty urban structures.

Hamilton dealt with this problem by building giant underground detention tanks that hold this combined sewer overflow and stormwater until some of the sediments settle out and the sewage treatment plant is ready to deal with the extra volume. Toronto has built two such tanks in the city's eastern end in the early to mid '90s, and Sado notes that there has been a measurable improvement in water quality in this area.

Sado's feeling – shared by many – is that the "mega problems" of the Toronto area demand "mega solutions," but others have their doubts about expanding rather than rethinking existing systems. Plans for a second set of tanks in the city's west end were continually delayed by citizen demands for closer scrutiny of its potential effectiveness. The project, however, was eventually approved and construction of the 4 km (two and a half mile) long system of storage tunnels and tanks should be completed in 2000.

A World Wildlife Fund (WWF) report puts the problem succinctly: "Taxpayers have invested billions of dollars in infrastructure which redistributes pollution rather than prevents it." What WWF researchers are concerned about are the toxins found in sewage systems such as Toronto's which are used by both industry and residents. Sewage plants have simply never been designed to deal with heavy metals such as copper and cadmium, or volatile organic compounds (VOCs) such as chlorobenzenes and xylene, that too frequently find their way into the city's drains.

Many metals attach themselves to other solids and settle out in the treatment process. But this is simply redistribution: if the resulting metal-laden sludge is incinerated, for example, some toxins end up going up the stack while other, now more concentrated toxins remain in the ash that is shipped out for landfilling.

Many of the volatile organics never even make it to the treatment plant – they evaporate along the way. And of the VOCs that do reach the end of the pipe, 73 percent, according to WWF, evaporate during treatment.

The only real solution to this problem is for industry to sharply reduce toxic discharges and eliminate discharges of persistent toxins altogether. The legislative motivation for industry to take these steps, however, has been slow in coming. The Province of Ontario's "Municipal Industrial Strategy for Abatement" (MISA) is years behind schedule in setting effluent standards.

But the culprit is not just industry. An estimated 15 percent of the toxins in urban systems come from households, in the form of everything from corrosive drain cleaners to the runoff from lawn pesticides and herbicides. It has been calculated, for example, that one-third of the insecticides used yearly in the Lake Ontario watershed are bought by urban residents, and that a total of more than 290,000 kg (640,000 pounds) of herbicides, insecticides, and fungicides are used around Lake Ontario each year by urban dwellers looking to keep some of nature's more inconvenient aspects at bay.

Such use is symptomatic of the historical indifference, or even hostility, shown by many people to natural processes. We have been unprepared in the past to accept that we live within a dynamic system. But that is slowly changing. When water levels in Lake Ontario hit a peak in the early 1970s, the call was for engineered solutions – dams, break-walls, gabions – to keep the lake in place. When the lake rose again in the early 1980s there was much greater acknowledgment that the solution was to stay out of harm's way – to build well back from shorelines and the tops of unstable bluffs.

The last decade has also seen an explosion of interest in rediscovering the wild in our cities. Ambitious efforts to rehabilitate wetlands, shorelines, and streams are being driven largely by citizen groups. Hundreds regularly turn out for tree plantings or garbage clean-ups. In Hamilton those working to restore the large marsh known as Cootes Paradise are waging battles on a number of fronts, from ensuring that cleaner water enters the marsh from feeder streams to blocking the entry of destructive non-native carp that uproot the carefully restored marsh vegetation. In Toronto citizen groups are working to bring life back to dozens of urban rivers and their tributaries: Black Creek, Emery Creek, the Humber, the Don and the Rouge Rivers, and even the long-buried Garrison Creek.

But as each group works on the problems of its individual watershed, there is also the recognition that real success will only come through addressing macro problems at the same time – from ill-considered urban development in headwater areas to stormwater runoff and toxic discharges.

A professional forester who has worked on projects in a number of Toronto's river valleys notes, for example, that because so little stormwater actually infiltrates the ground in modern cities, urban valley bottoms which were once rich, moist habitats are now bone dry. "We're planting cedar, dogwood, cottonwood and other traditional riverine species" – species which are adapted to life on the wet edge of a river, he says, but "whether they will 'take' under these dry conditions, I really can't say."

"We do have to start treating stormwater as a resource," affirms Ed Sado. Stormwater, in other words, can become the first test of our urban commitment to start working with natural processes rather than against them. And Sado does see a new attitude evolving: Etobicoke in western Toronto, for instance, has introduced an innovative policy of using porous storm drains that allow for natural infiltration when replacing older interconnected pipes. Other municipalities are unhooking residential downspouts from their stormwater systems, and they are encouraging residents to use less water – thereby reducing the number of potential overflow situations. In many new developments, stormwater management is an integral part of the planning process.

For David Crombie, former Toronto mayor and leader of the Royal Commission on the Future of Toronto's Waterfront, and now of the Waterfront Regeneration Trust, working with nature is a road which cities simply must go down. "This is the path to survival for cities," he says. Cities are clearly the human habitation of choice, but only by integrating an understanding and respect for nature into urban living will they remain liveable. He adds, "We need nature in the city because it teaches us respect, and that is the single most important message we can get out today – how everything is interconnected and how that means we must have respect for other entities and other communities" – both human and natural.

# Stories from a Big City Stream

## *Restoring Toronto's Don River*

*Gregor Gilpin Beck*

On the day of this flood (facing page), volunteers were preparing to plant thousands of aquatic plants in a newly created habitat wetland. Unfortunately, heavy rains caused a flash flood that submerged the lower Don recreation trail and postponed the event. To be successful, restoration efforts must address the entire watershed. For the Don River, that means sound protection – from the sensitive Oak Ridges Moraine headwaters, all the way downstream to Lake Ontario.

RAIN STREAMS RELENTLESSLY against the windowpane of my Toronto home as sparse traces of an exceptionally mild winter are washed away. Once again the temperature belies this January date. While a walk in the ravine is an enjoyable diversion from winter in the city, an oppressive heaviness hangs in the air. Footing on the trail is treacherous: slick, raw earth is exposed on hill slopes denuded of vegetation, and vapours steam from sheets of ice along this small tributary of the Don River.

Short weeks earlier, with a northern chill and fresh snow hushing the city, the flow in the stream was minimal. The air seemed fresh, the stream clear and clean, and escape was near complete. With recent torrents, the creek's volume has surged. Its colour has changed, the clarity obscured. Dankness and quiet winds blend with the rank odour of the water on this day.

Unseen and underground for most of its brief course, the creek emerges into the wild and tangled ravine where as a child I played with homemade boats and delighted in small discoveries. Further downstream now, I see the water disappear into lifeless darkness of metal and concrete tunnels and

Facing page: Flooded trail, Don River. Photo: Gregor G. Beck

culverts, then re-emerge to linear channels and angular gabion banks. In too few places, it flows free.

Several months later, I am looking at the Don River from a different and perhaps more appropriate perspective. Having walked, cycled, and skied through these ravines throughout my life, I am now riding the modest swells of the main river's middle reaches in my weathered cedar-strip canoe. I feel that I know the river well, yet I am surprised by the increased intimacy of this relationship when travelling as an integral part of the living flow. I am surprised, too, that despite the obvious pollution and habitat destruction, this humble urban experience is so truly enjoyable.



There *is* intrigue in this voyage. As with river trips everywhere, one wonders what is around the next bend. Will I surprise a belted kingfisher searching for minnows in a placid pool? Perhaps glimpse another hooded merganser resting on a log overtop a shimmering riffle, or maybe a turtle, or even a beaver farther downstream? Or will I come face to face with the grim smile of an iron retaining wall below a six lane highway? Regardless of the intrusions and the inescapable din of traffic, it is nourishing to the spirit to see curving riverbanks, soft and free with vegetation, and not just a concrete and steel trench confining a strait-jacketed prisoner.

I have planned the timing of this urban canoe trip carefully. One must get on the river fairly soon after a rain to ensure sufficient water for passage. But too soon, and the river is dangerously treacherous; too late, and the flow is pathetic.

Most of the Don River watershed is covered with roof tops, asphalt, driveways, mowed lawns – and precious little natural forest cover. Melting snow and rainwater are whisked away through sewers and culverts before there is time for absorption by soil and plants. The result is an artificially dry riverine habitat and a watercourse that experiences periodic flash floods. Sadly, water is far too often treated as a nuisance and not as a valuable resource.

Most of the time my small tributary stream is only inches deep and mere feet across. However, on regular occasions it is an angry torrent of so-called stormwater, complete with not just rainfall but also tainted runoff from the city's hard surfaces and sewage which might otherwise overload treatment facilities. I have seen the titanic alter ego of this creek. It becomes a raging menace, at times six feet deep and much greater than that in width. Sewer covers upwards of 150 pounds can pop as if they were corks. Like its tributaries, the main Don River overflows its banks too, and sometimes drowns those who challenge its might. During these times, the river can silence the hum of traffic by flooding the valley's roads.

The Don watershed shows the sorry signs of wear from over two hundred years of settlement and from its 800,000 residents. Eighty percent of the watershed is now urbanized, up from 15 percent as recently as 1950; the Greater Toronto Area is home to nearly five million people.

Regardless of their location, urban rivers suffer grievous environmental insult. What were once tremendously important natural wildlife corridors have become highly developed human corridors and major centres of population. Nonetheless, Canada's largest metropolis is fortunate to have the extensive green space of not only the Don but several different river systems, each with a diverse and relatively abundant flora and fauna. And throughout the Don's watershed, extensive efforts are underway to improve the quality of the river and surrounding habitats and to attempt to undo some of the historic damage.

My first recollection of trying to clean up the river goes back to early school days. Twenty-five years later I still recall images of small children in rain gear picking up garbage from "my" small stream and the ravine slopes adjacent to the school. Conservation efforts have also focused on restoring tens of thousands of native trees, shrubs, and wildflowers. The task is simple and extraordinarily effective, and continues to the present. More vegetation means more habitat for wildlife, and the other spin-offs are seemingly endless. Plants consume carbon dioxide, thereby reducing greenhouse gas emissions; they retain water which reduces flooding and sewer overflows; they help prevent erosion on river banks; and by shading rivers, they keep the water cool, which increases the amount of oxygen for fish and other aquatic creatures. More recently, planting projects have been initiated in concert with attempts to remove alien (or non-native) species that frequently take over large areas and decrease the land's usefulness for native wildlife.

Other projects are more complex and require careful planning and public consultation, exacting ecological execution, and a lot of fund-raising. Several wetland restoration projects are being completed in Toronto to address the loss of so many of our natural marshes and swamps. These projects will recreate habitat, educate the public about the importance of aquatic ecosystems, and in some cases actually help to reduce the amount of pollutants and excess nutrients entering the Don River. The names of wetland restoration projects include "Chester Springs," "Todmorden Mills Ox-bow," "Heron's Roost," and the "Don Brickworks" – a celebration of the ties that bind human history with natural history.



Small but significant improvements are being made in an effort to re-establish the natural wildlife corridors that exist along rivers. Throughout history humans have attempted to control rivers with dams and weirs. The associated environmental problems are numerous, including the blocking of migration routes for fish. Two weirs on the lower Don have been modified now to improve this situation. Flood control and road stability concerns meant that the weirs could not be removed entirely, but a series of rock pools has been created on the downstream side, reducing the vertical drop sufficiently to allow fish such as trout, bass, and pike passage further up the river. Thoughtful design considerations hope to create spawning habitat for walleye and block the movement of the non-native sea lamprey. More weir modifications are planned for further upstream.

One of the most important things being done in the Don is to publicize the importance of protecting and improving our urban waters. Public education projects include presentations to schools and interested groups, interpretive signs and walks celebrating the watershed, and even door-to-door campaigns. Among the public education programs are those that encourage people to use water wisely and respectfully. If we all reduce our water consumption a little, then there will be fewer water pollution problems and less need for expensive engineered solutions to stormwater and sewer water treatment. In parks and green spaces across the city, improved ravine access and extensive trail systems will encourage greater public interest and concern about our environment, and ultimately greater stewardship.

The restoration challenge is perhaps greatest at the highly degraded downstream end of the river. Plans continue to evolve in an attempt to turn the green corridor printed on the City's Official Plan into something ecologically meaningful. Recently, discussions have begun regarding water quality in Toronto Bay, formerly known as the inner harbour, and politicians from all levels of govenments are beginning to show their support for the "greening" of Toronto's waterfront – at least in principle. The task is difficult in these very old parts of the city where existing infrastructure and buildings make any change difficult. Will we ever see the original "marsh at the mouth" restored? Will small animals ever be able to travel north from Lake Ontario through a winding, forested river valley to the headwaters on the Oak Ridges Moraine? When will we be able to remove the signs on the Don that state so sadly, "Polluted Waters: No Swimming, No Wading"?

Fortunately, a great many groups and agencies are involved in trying to improve the Don River and its watershed. Among them are Toronto's citizen-based Task Force to Bring Back the Don, Friends of the Don East, and the Don Watershed Council. Other agencies include the Waterfront Regeneration Trust, the Toronto Bay Initiative, Toronto Region Conservation Authority, and Toronto Parks and Recreation.

At meetings around the watershed we sometimes clash on how best to achieve our conservation objectives. Sometimes, too, we seem to focus too much on bureaucratic planning processes and too little on the ecological results that we seek. In the end, though, we differ little on the fundamentals of what could and should be done to improve our environment. And on the bright side, of course, any and all restoration achievements represent steps in the right direction.

The challenge to restore the Don River's water quality and terrestrial and aquatic habitats is shared in urban watersheds everywhere. The history of human settlement and the resulting environmental degradation echo through the many cities around the Great Lakes and the St Lawrence. One need only recall that both Toronto's Don River and Cleveland's Cuyahoga River were once so polluted that they actually caught fire. Through the tremendous efforts of many, Toronto and dozens of other cities are working hard to ensure that historical chapters of this sort remain firmly in the past.



Nearly every day I return to wooded trails for my habitual walk beside a little creek. Regardless of the smell, regardless of the clarity, regardless of the unseen road salt, chlorine, poisons, and wastes, this small stream continues to flow. It endures to record its stories of the land, the air, and the people around it. Here in the roadway-infested valleys of Toronto, the Don River swells with stories from all of its many tributaries. And like waterways everywhere, it shares them with those who care to listen.

*"Most of the Don River watershed is covered with roof tops, asphalt, driveways, mowed lawns – and precious little natural forest cover. Melting snow and rainwater are whisked away through sewers and culverts before there is time for absorption by soil and plants. The result is an artificially dry riverine habitat and a watercourse that experiences period flash floods. Sadly, water is far too often treated as a nuisance and not as a valuable resource."*

Photo: Frozenrope

Photo: Gregor G. Beck

# Decline and Recovery

## *Restoration of Hamilton Harbour*

*Louise Knox*

*"It takes a long time for an ecosystem to recover, but Hamilton Harbour may be the place to demonstrate that it can be done, especially if all stakeholders continue to cooperate. For those of us who follow poetry, a slight change of tone is emerging, with a return to the sense of awe inspired by nature in some of the local poetry around the harbour. Let's hope this trend deepens and extends to many more communities in the future."*

IN THE LATE 1960S AND EARLY 1970S, as the industrial world was awakening to the damage it was doing to the environment, I was studying twentieth-century poetry. Most of what I read conveyed the alienation and loneliness of poets who did not seem to feel connected to people, or to nature, or to any spirit outside their own. Their work reflected a difficult era in western history, during which two world wars and very rapid industrialization had destabilized communities and caused more pain than most people were inclined to recognize.

This was the era in which ports and harbours around the Great Lakes and the St Lawrence suffered tremendous environmental damage. By the 1960s, if the poets had not been eloquent enough to reach people's sensibilities, the environment would have spoken loudly enough for itself. The Cuyahoga River and the Don caught fire. Lake Erie was declared "dead." Rachel Carson wrote *Silent Spring*.

Disturbing as these events were, they caused an awakening, and the second half of the twentieth century has been a time for rebuilding and restoration. The Hamilton Harbour story illustrates one such transition from decline towards recovery.

Hamilton Harbour is a largely enclosed embayment at the western end of Lake Ontario, separated from the lake by a narrow sandbar with a canal cut through it. It is 2,150 hectares (5,300 acres) in area with a drainage basin about ten times that size. A narrow strip of land separates the harbour from a large marsh called Cootes Paradise.

Prior to 1900 the harbour was renowned for its thriving fishery and expansive wetlands. In the twentieth century, it became home to what was once the largest concentration of heavy industry in Canada. For a fledgling country, the steel manufacturing capacity was seen as essential to political and economic sovereignty. Local residents, many of whom were new immigrants at the time, believed they could either have jobs and a strong manufacturing economy, or they could swim in the harbour – but they could not do both. They also believed that negative impacts on the environment would be limited to their local community.

During the 1960s these beliefs were challenged. Scientists began to see and describe impacts on the environment that were distant from the sources of contaminants, both in space and in time. Hamilton became known throughout the country and in the United States as a pollution hotspot. Residents became ashamed of the look and smell of the harbour, and they resolved to do better. Fortunately, pollution control technologies began to surface.

From the mid 1970s to the mid 1980s local resolve was aided by federal and provincial regulations. These combined forces brought about reductions in pollutant loadings from the industrial sector of between 80 and 95 percent for a number of toxic substances. Reductions in phosphorus, ammonia, and suspended solids were achieved in the municipal sector through waste-water treatment plant improvements.

In the mid 1980s, the Canadian and U.S. federal governments amended the Great Lakes Water Quality Agreement. This document committed both nations to developing Remedial Action Plans (RAPs) in Areas of Concern on the Great Lakes, one of which was Hamilton Harbour. The oily sheen and bad smells were no longer evident, but this agreement recognized that the full restoration of the areas was still not assured. A requirement in developing these plans was that they be written in consultation with the provinces or states, and also the public.

By 1992 a plan had been finalized for Hamilton Harbour based on a broad consensus of all stakeholders (environmental, industrial, governmental, academic) and the community at large. It would attempt to balance industrial, commercial, and recreational interests, including swimming in some areas.

The Remedial Action Plan called for a reduction in the number of occasions during which raw sewage combined with stormwater would overflow during and after heavy rainfalls. This was to be accomplished by building twelve storage tanks and one tunnel that could hold back the volume of effluent during the storm event and then channel it to the sewage treatment plant during low flow times. By 1997 four such tanks had been built, and others were in the design phase. Strategically located in the harbour's western end, these tanks made the beaches swimmable during most of the summer at the new Bayfront Park.

The plan called for increased public access to the harbour shoreline. By 1993 contaminated soil in an old industrial dump had been cleaned up and Bayfront Park was created. By 1997 the amount of publicly accessible shoreline had climbed from less than 5 percent to an estimated 16 percent.

Restoration of fish and wildlife habitat was also recommended. More than 65 percent of the shoreline habitat had been lost to industrial and port development prior to 1970 and could not be recovered. Other areas, such as the Cootes Paradise marsh, provided opportunities for habitat restoration or the creation of new habitat. By the end of 1997 various partners had contributed over $13 million to create about 63 hectares (155 acres) of this new habitat, which included islands, trails, and viewing stations, and opportunities for scientists to study restoration ecology.

Excessive algae and poor water clarity remained problematic. The plan sought further reductions in phosphorus, ammonia, and suspended solids coming from various sources. To help address this, the operation of the Skyway Waste Water Treatment Plant in Burlington was optimized in the mid 1990s. By 1997 the plant was achieving

Eastern shore, Georgian Bay. Photo: Gregor G. Beck

the target for phosphorus fairly reliably. Whether this could be sustained while lowering the ammonia loading remained to be seen. Improvements at the Woodward Treatment Plant were also recommended, as were improvements in erosion control further upstream within the watershed.

Toxic pollutants in bottom sediments remain problematic because they recirculate into the water column and contaminate wildlife. This ultimately restricts the consumption of certain sizes and species of fish. The RAP recommended investigation and early action to remediate these sediments, but this has proven to be an area where action is difficult to facilitate. Despite developing new technologies and completing several investigations of an area near Randle Reef adjacent to the Stelco docks, little progress had been achieved by the end of 1997.

Most of the actions recommended by the RAP were followed voluntarily by the agencies and organizations which sponsored them. This attests to the importance of community involvement in the planning and implementation phases of the process. Community stakeholders have formed an organization called the Bay Area Restoration Council (BARC) to monitor the implementation of the plan and to ensure that the public is involved and informed as it proceeds. BARC continues to create incentives for organizations to act by reminding them of community expectations.

It takes a long time for an ecosystem to recover, but Hamilton Harbour may be the place to demonstrate that it can be done, especially if all stakeholders continue to cooperate. For those of us who follow poetry, a slight change of tone is emerging, with a return to the sense of awe inspired by nature in some of the local poetry around the harbour. Let's hope this trend deepens and extends to many more communities in the future.

Thousand Islands, St Lawrence River. Photo: Frozenrope

# The St Lawrence River

Part of our environmental challenge is to overcome the perception that the Great Lakes and the St Lawrence River are separate systems. They are, of course, part of one vast and ecologically intertwined watershed, and the health of the St Lawrence depends on the health of the Great Lakes, and vice versa. Perhaps a simple and singular name for the entire watershed might help to overcome this misconception. Louis-Gilles Françoeur has suggested the Laurentian Basin – or, "le Bassin laurentian." Regardless of the name, there will always be a need to treat the watershed as a whole, not as separate component parts.

Photo: Frozenrope

# Ecosystem in Peril

## *The St Lawrence River*

*Louis-Gilles Françoeur*

*"Throughout the modern history of the St Lawrence, economic considerations have always taken precedence over the survival of wildlife and the river's ecological well-being. It is necessary to reverse our order of priorities and allow the survival of this great ecosystem before it ends up as the Great Lakes sewer, glorified as a maritime highway. We rely on the St Lawrence, although we continue to neglect it. The challenge now is to make the fate of this vast watershed a central part of our collective vision for the future."*

THE HISTORY OF THE ST LAWRENCE RIVER began 60 million years ago, when the same tremendous geological forces that separated the continents carved out the river valley. The geography of the region as we now know it, however, was shaped by glaciers just seven thousand years ago and emerged with the retreat of the Wisconsin Ice Sheet and the Champlain Sea.

For 6,500 years the people who made their home along this 750 kilometre (450 mile) river lived in harmony with the other species in this 840,000 square kilometre (325,000 square mile) ecosystem – an ecosystem comprised of one of the biggest rivers in the world, fed by lakes which contain about 20 percent of the planet's fresh water, and flowing into an estuary the size of a sea. But with the arrival of white settlers and the subsequent growth of industries, the relationship between humans and environment changed profoundly.

Although it was not immediately obvious, the building of towns at the mouths of tributaries significantly affected the ecosystem, since many of these locations bordered vital fish spawning and feeding grounds. Initially, the abundance of fish was such that even the harm caused by draining wetlands and floodplains for farms or road construction did not produce a noticeable decline in fish stocks. However, with the industrial revolution, the scope of development broadened considerably and the changes to the river's ecosystem became visible. From this period on, cities swelled in size and dams were built to meet growing energy needs. Shipping, which had already allowed the colonization of an otherwise inaccessible region, was transformed by the invention of the steam engine. As a result of these and other changes, migration routes of certain fish were blocked, new species were introduced, and the flow and habitats of the river were profoundly altered.



## The Decline of an Enormous Ecosystem

The expansion of the shipping industry which followed the introduction of steam power transformed Montreal into a leading commercial and industrial centre, since heavy steamships were restricted from travelling further west by the rapids just outside the city. The Lachine Canal, completed in 1875, opened the Outaouais (Ottawa) River region to smaller ships, and heavy industry developed along its banks. Today the canal is one of the St Lawrence's biggest sources of toxic waste.

After the Second World War a coalition of powerful business groups pushed for improved shipping conditions between Montreal and the Great Lakes. Ottawa eventually agreed and authorized the construction of the St Lawrence Seaway, a decision that would eliminate Montreal's strategic geographical advantage and its place as Canada's largest metropolis. With the completion of the project in 1959, it became possible for 25,000 metric tonne ocean-going ships to reach the Great Lakes through the new locks at St Lambert and Beauharnois. Today two thousand such ships are using the St Lawrence Seaway each year. This massive undertaking opened a shipping route more than 3,000 kilometres (1,860 miles) long into the heart of the North America continent.

The St Lawrence navigable waterway also includes the Saint-Laurent Channel, two hundred metres wide and eleven metres deep, which has significantly changed the river's aquatic life. Between Montreal and Quebec City, it monopolizes between 30 to 50 percent of the water in the river. During July as much as 90 percent of the river's flow at Montreal goes into this channel. On the surface nothing looks out of the ordinary. However, the environmental impacts of this enormous creation (from which a million cubic metres of sediment must be dredged every year) has yet to be assessed through the unbiased eyes of independent researchers.

According to a 1995 federal government study, dredging the Seaway every year has little effect on the river. The same document, however, admits that "some researchers" blame the procedure for a decrease in certain fish stocks – a theory, the study is quick to point out, that has never been verified. Many anglers from around Lac Saint-Pierre (between Montreal and Trois-Rivières) report that currents along sections of the river in that area have been slowed. In addition, water levels have been artificially raised in the centre of the river following the construction of small dams between various islands; as a result, fish habitat has been drastically changed.

Not only does the Seaway take much of the river's volume but it also increases the speed of some currents. According to biologists, the result is a veritable wall of water that many fish species cannot cross. Such conditions diminish the biological exchange between banks, possibly changing the routes of migratory fish species that would otherwise use both banks. One of the few studies to address this question noted major changes in the geographic distribution of Atlantic sturgeon, American shad, tomcod, American eel, and striped bass. The latter disappeared from the river after the Seaway was built.

Fishing and conservation groups have begun to assess the impact on wildlife of the thousands of ships and boats. In 1991 there were 10,461 shipping trips made up and down the Saint-Laurent Channel, as well as trips by 867 passenger vessels and thousands of leisure craft. Among many environmental problems is the erosion of river banks by waves created from large ships and the disturbance of bottom sediments. Meanwhile, the damage caused by heavy shipping has led governments, municipalities, and private citizens to reinforce large stretches of river bank with rock fill that is as ugly as it is unproductive for river wildlife. In 1992 the federal government permitted the

Seaway to increase its depth by a foot, and further dredging was authorized early in 1998 to accommodate even larger shipments. Ecologists are asking if or when the ecosystemic threshold for the system will be reached – but nobody seems to have the answer.

With the present levels of shipping traffic on the St Lawrence, the risk of accidental spills is very high. Between 1971 and 1988, 641 spills were reported, though fortunately most were minor. Still, no less than 21 percent of the material transported on the river poses a potential threat to the aquatic environment.



Regulating the water flow in the river and its main tributaries to control flooding, maintain adequate depths for shipping, and generate electricity has had further negative impacts on wildlife. The controlling of water levels in the Great Lakes has reduced the threat of serious flooding for people living next to the river, but without high waters, certain fish species have been denied access to long-established spawning grounds in flooded fields and forests close to the river's banks.

Major changes to the hydraulics of the river have also been caused by hydroelectric development to the river and its tributaries. New York State, Ontario, and Quebec have all dammed the river in places, cutting off migration routes of various fish species and aggravating changes in seasonal water cycles. These natural rhythms have been greatly disturbed by the construction of hydroelectric dams on tributaries as well, such as the Outaouais (or Ottawa), Sainte Marie, Saguenay, Outardes and Manicouagan, and most recently, the Ste Marguerite.

The species most affected by these developments are the fishes that live in the sea and swim upstream to spawn in fresh water (i.e., anadromous species). These include the American shad, mooneye, and Atlantic salmon. The American eel, which does a reverse migration, is also affected.

Two additional factors complete this picture of an ecosystem in peril: the destruction of marshes and the reshaping of tributaries. During the 1970s and 1980s, while Quebec spent $7 billion cleaning up urban sewage, developers and farmers were allowed to destroy vast areas of wetlands.

Between 1945 and 1976, approximately 3,650 hectares (nine thousand acres) of shoreline wetlands on the St Lawrence were wiped out. Thirty-four percent of this loss was from agricultural development. And these numbers do not take into account the damage caused by the construction of towns at the mouths of tributaries over the preceding three hundred years. Between 1945 and 1988 11,500 hectares (4,650 acres) of the river bed was dredged, 55 percent of this from deep waters (mostly in the ship channel), having a large impact on bottom-dwelling (benthic) animals and plant communities.



Montreal has contributed more to the destruction of wetlands along the river than any other city in Quebec, allowing real-estate developers to obliterate five thousand hectares (two thousand acres) in twenty-five years. The construction of the islands for Expo 67 was one of the single most damaging projects. With the loss of 80 percent of its marshes, Montreal's riverside or riparian habitats are the most artificial of any city on the St Lawrence.

Some 43,000 hectares (17,400 acres) of these riverside wetlands remain: 32,000 hectares on Lac Saint-Pierre and eleven thousand at Lac Saint-Francois. By the year 2000, the federal government will ensure the protection of ten thousand hectares of this land as part of its action plan for the St Lawrence River.

An evaluation has yet to be made to measure the effects of reshaping the tributaries that feed the river. Between 1960 and 1990, to speed water drainage from farmland in the spring and to lengthen the growing season, the Quebec government financed the straightening and deepening of twenty thousand km (12,400 miles) of watercourses in agricultural areas.

During this period, countless streams and small rivers – many of which serve as "nurseries" for various fish species – were transformed into unsightly ditches which have become sediment traps and dumping grounds for toxic pesticides and fertilizers. One such watercourse, the small Rivière Boyer in the Lower St Lawrence region, had in fact been the biggest spawning ground on the south shore of the gulf for smelt. This small fish is one of the region's principal marine forage fish for larger species.

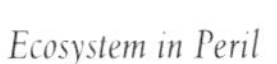

### Pollution

The 250,000 metric tonnes of contaminants which are spewed into the river by industries and towns are particularly damaging to fish species whose habitats are disappearing. It is important to note that 42 percent of the toxic pollution in the St Lawrence River originates in the Great Lakes. However, Quebec's cross-border pollution problem appears to be largely ignored by Ontario and the United States, which together are also responsible for 75 percent of the acid rain and 60 percent of the smog hitting the French province.



In the past the Great Lakes have been known to pollute the river with PCBs and pesticides such as DDT and mirex, as well as heavy metals, primarily arsenic, selenium, and mercury. To this Quebec polluters add chlordane, pyrene, and chlorobenzenes. They also dump most of the heavy metals associated with the chemical and metallurgical industries in proportions similar to those found in the inflow from the Great Lakes. The PCB concentrations are higher around Montreal, while the polycyclic aromatic hydrocarbons (PAH) appear to come mostly from around Quebec City, even if there are no known sources for PAHs in that area. Some researchers think the material is introduced into the environment as atmospheric fallout.

Pollution from the Great Lakes tends to keep to the middle of the river, which is a concern for human health. About 45 percent of Quebecers get their drinking water from the river, with intakes generally as far from shore as possible to avoid local pollution problems.

A couple of miles upstream from Quebec's western border, in Massena, New York, the river banks next to factories belonging to General Motors, Alco, and Reynolds contain up to 84,800 tonnes of contaminated sediments. Concentrations go up to five thousand parts per million of PCBs. Every day more of these old deposits of PCBs ooze into the St Lawrence and Lac Saint-François. Fish species that feed on smaller fishes in this lake show high levels of heavy metals, pesticides, and pollutants even if there is not a single industry on its banks.

Most of the toxins that enter the river in Quebec are dumped by the region's major chemical and metallurgical companies, located on the south shore between Cornwall and Sorel at the entrance to Lac Saint-Pierre. Quebec's oil refineries are concentrated on the island of Montreal.

The high levels of PAH and organochlorine in the river may have contributed to the scarcity of certain fish species such as sturgeon. The copper redhorse, a fish species now found only in the Chambly rapids on the Rivière Richelieu, is close to extinction. The males contain dangerously high levels of pesticides, perhaps explained by their diet of snails and mussels, both of which are know to bioaccumulate toxins.



The eels found in Quebec, their migration route blocked by the dams and locks of the Seaway, were severely contaminated by mirex, a pesticide originally used against red ants in the southern United States. It was produced and dumped into rivers upstream in New York State. Farther up the food chain, the endangered St Lawrence population of beluga whale accumulates very high levels of mirex, and dozens of other contaminants.

Despite all the bad news, Quebec has made great strides since the mid 1980s to control pollution in its section of the St Lawrence River. By 1996 toxic pollution generated by the fifty biggest polluters, including major paper producers, had been reduced by 96 percent compared to 1988 levels. At that time they were cumulatively dumping 574 tonnes of suspended solids, creating a 446-tonne biochemical oxygen demand (thus reducing the precious oxygen content of the water), plus 1.8 tonnes of oil and grease, 0.9 tonnes of heavy metals, and 75.5 tonnes of other metals such as iron, aluminum, and vanadium. The 4 percent that remains from the original toxic discharge from the fifty biggest polluters is equivalent to the non-treated pollution discharge from fifty-six other, newer large industries on St Lawrence tributaries. These are targeted by phase two of the federal-provincial pollution control program.

During the 1980s Quebec treated only 2 percent of the waste handled by municipal sewer systems. By the year 2000, 98 percent of this was to be treated by modern plants equipped with secondary treatment. Some of these plants are almost as efficient as tertiary treatment facilities, and Quebec may be able to export new technologies.

The St Lawrence River receives a great amount of pollution from its tributaries. A 1993 joint federal-provincial study revealed that the tributaries collectively contribute over three times more pollution than the industries on the St Lawrence itself. This is partially explained by the fact that the tributaries are home to some major polluters, including paper-making industries. Once the foremost polluters in Quebec, these are now among the greener large corporations. The watersheds of the Outaouais and Saint Maurice rivers have been heavily deforested for agriculture and paper production. They belong to a group of the six most polluted tributaries of the St Lawrence. The others are the Richelieu, Saint-Francois, Batiscan, and Yamaska rivers, all of which flow through southern agricultural lands, the main source of non-point pollution.



The ecosystem is also affected by everything that happens within the St Lawrence River's enormous drainage basin: increases in pollution levels, changes to natural habitats, dam construction, deforestation, the heating of agricultural waterways, and erosion.

## Diagnosis

The reductions in pollution achieved through the St Lawrence action plan are a sign of hope. But despite short-term improvements to the quality of life for people near the river, the ecosystem as a whole does not appear to be recovering.

A unique historical picture of the relative abundance of fish species in the St Lawrence is provided by the Quebec aquarium. Since the early 1970s this government institution has fished the river to replace and feed the species it displays – always fishing in the same place on the river and using the same equipment. In 1971 the aquarium netted fifty thousand fish. In the last five years it has caught fewer than ten thousand in all. It seems unbelievable that the biological productivity of the river could have decreased by 80 percent in less than a generation for many species. Unfortunately, no comprehensive studies are available which would help to validate this important hypothesis.

The most threatened species include eel, tomcod, shad, smelt, and yellow perch, although the latter is still found in great abundance in the Montreal region. These species, which have been practically wiped out in certain sections of the river, are being replaced by walleye, shorthead redhorse, bass, and lake whitefish. Other studies have shown similar declines. An 80 percent decrease in eel at Cornwall supports the data from the aquarium.

Another threat faces the fish of the St Lawrence. This one is biological in nature, namely the introduction of exotic species. Zebra mussels and the sea lamprey have had a tremendous impact on aquatic life further upstream in the Great Lakes. The introduction there of rainbow trout and several Pacific salmon species for sport fishing has further changed nature's balance. These species compete with native trout and salmon family fish for food and habitat. The seemingly carefree way in which these species are stocked, without even a cursory impact assessment, is alarming. In contrast, the rainbow trout introduced to new habitats in Quebec for sports fishing are sterilized, a preventive strategy that should be followed in the Great Lakes.



### The Future

Contaminants have flooded into the St Lawrence from the Great Lakes and atmospheric fallout for years, and they will likely continue to do so for a long time to come, despite the many clean-up projects underway. The measures described above to reduce pollution are important not only for the river but for sensitive sections of the river's vast estuary and the oceans. Pollution in the river reaches the long marine trench that starts near Tadoussac, at the confluence of the Saguenay and St Lawrence. Heavy metals and even dioxins have been detected in this very deep part of the river, albeit in very low concentrations.

In 1827 Montreal introduced a regulation requiring citizens to bring their garbage to a public beach where it was disposed of by being thrown into the river. This established an approach to waste and river management which is clearly not sustainable or responsible. Witness the 80 percent reduction in some fish stocks near Quebec and the loss of a thriving commercial fishing fleet.

Some researchers are now saying that the river cannot be brought back to life without a strategy for restoring its vital ecological functions. Such a strategy would begin by ensuring complete protection for all spawning grounds and wetlands that have survived the short-sighted development of recent decades, as well as the restoration of vital key habitats that have been lost. However, given its international scope, the second step could represent an unprecedented political and scientific challenge.

To re-establish the river's normal circulation patterns it would be necessary to carry out a thorough impact assessment. This could lead to questions on the future of Seaway operations (if not on the future of the Seaway itself) that have fundamentally altered the hydraulic flow of the river and changed the migration and reproduction patterns of so many fish species. Ontario and the United States will have to face up to their responsibilities in this debate, since it is primarily their interests that have been served by turning North America's biological artery into a shipping highway.



An analysis must also be made of the impact of the greenhouse effect on land next to Laurentian rivers, the Great Lakes, the St Lawrence River, and indeed all coastal areas. The geography of this vast watershed was formed by large-scale climatic changes during the last glacial period. I believe that greenhouse gases and global warming threaten to change the river's ecosystem more profoundly than anything since these transformations of long ago.

The Canadian Climate Centre predicts that water levels in the Great Lakes will decrease by 30 to 80 cm (one to 2 1/2 feet) and possibly more in the next one hundred years, resulting in a 20 to 50 percent reduction in stream flow and also the possibility of no seasonal ice. This would create profound changes for many winter migrants in the St Lawrence, such as the tomcod. Reduced streamflow would also concentrate pollutants, and stronger currents could harm spawning grounds and cause erosion.

Throughout the modern history of the St Lawrence, economic considerations have always taken precedence over the survival of wildlife and the river's ecological well-being. It is necessary to reverse our order of priorities and allow the survival of this great ecosystem before it ends up as the Great Lakes sewer, glorified as a maritime highway. We rely on the St Lawrence, although we continue to neglect it. The challenge now is to make the fate of this vast watershed a central part of our collective vision for the future.

Photo: Gregor G. Beck

# Environmental Sleuths

## *Biological Indicator Species Help Track Pollution*

*Jean Rodrigue, Louise Champoux, and Jean Luc DesGranges*

Toxic pollutants accumulate in the tissues of aquatic life. This is particularly true for animals at the top of the food chain – or for those that filter-feed tiny plankton from the water, such as clams and mussels. Many parts of the St Lawrence remain closed to shellfish gathering, especially in areas close to urban centres, or downstream from heavy industry. In some cases, scientists are using tissue samples from these animals to monitor environmental clean-ups.

MORE THAN THIRTY THOUSAND chemical compounds are in use within the Great Lakes–St Lawrence River watershed. Eight hundred are considered toxic, and many of these are known to be highly persistent. Environmental improvements have been made, however, on many fronts recently: better production methods and safer chemical compositions, increasingly widespread treatment of waste-water, the banning of highly toxic pesticides such as DDT and dieldrin in the early 1970s, and tighter regulation of PCBs. Nonetheless, there are still traces of heavy metals, organochlorine pesticides, and PCBs in the waters, sediments, plants, and animals throughout the watershed – all the way downstream to the mouth of the St Lawrence and beyond. These contaminants enter and accumulate in the Great Lakes–St Lawrence system from agricultural runoff and urban waste-water, industrial plants, hazardous waste discharges, and the use of toxic pesticides, as well as atmospheric transport and fallout. Individually and collectively these contaminants are a threat to humans and wildlife alike.

The St Lawrence Vision 2000 plan is aiming for a 90 percent

reduction in the pollutant load discharged into the river from 106 priority industrial plants or "point sources." To achieve this goal, we need to develop a tool for measuring the success of these clean-up activities. One method is to assess how wildlife can answer questions about environmental problems. The Canadian Wildlife Service is evaluating the potential of amphibians, reptiles, and birds living in the region as pollution monitors or so-called bio-indicator species. By combining the study of factors such as diet, habitat, and exposure to pollutants with detailed chemical analyses, we will be able to identify which sites are most polluted and also chart improvements resulting from the clean-up programs.



Among the animals under study are two species that eat invertebrates and live near point sources of contamination. The mudpuppy (an amphibian) feeds on bottom-dwelling (benthic) organisms, whereas tree swallows feed on flying insects; each reflects somewhat different environmental conditions. The domestic Peking duck is used in controlled studies on local contaminants. Several other species are being evaluated for their usefulness as indicators of regional contamination. The common eider, a mollusc-eating duck, provides indirect information about the contamination of the suspended matter filtered by its prey. The snapping turtle, common tern, herring gull, double-crested cormorant, black-crowned night heron, and great blue heron can be studied to monitor contamination in free-swimming or nektonic prey such as fishes of various ages, sizes and diets.

Results from studies on mudpuppies and snapping-turtle eggs from the Ottawa and St Lawrence Rivers have shown noticeable geographical variations in contamination patterns, as well as other significant differences in contamination levels. For example, the PCB composition found in mudpuppies resembled the patterns and trends reported for fish, whereas contaminants found in turtles were more akin to patterns found in bird species.

The great blue heron and the black-crowned night heron have been studied in detail to select biochemical indicators that affect their health. These biomarkers are biological responses to chemicals that measure exposure and sometimes, too, the toxic effects. It is assumed that a healthy individual organism exposed to pollutants will show progressive deterioration in health, beginning with early initiation of biochemical or compensating responses. When the pollutant load gets too high, the organism cannot compensate, and the pathological or disease process begins. To be useful, a biomarker should produce a response that can be related to an easily identifiable degree of impairment of growth or reproduction which ultimately affects the survival of the individual or the entire population.



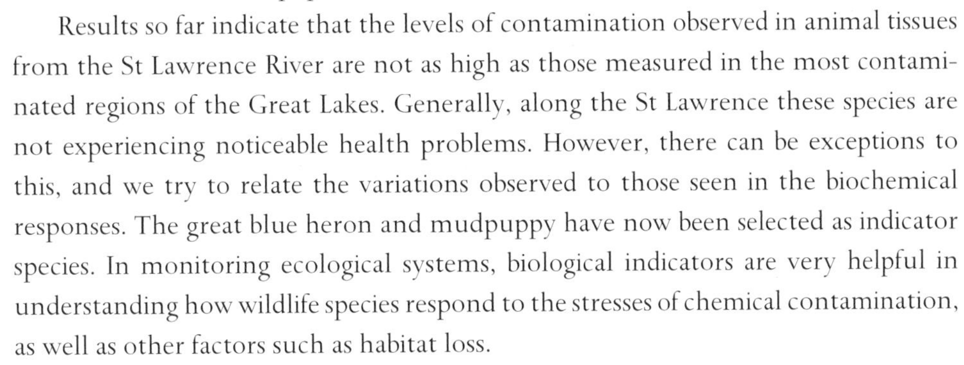

Results so far indicate that the levels of contamination observed in animal tissues from the St Lawrence River are not as high as those measured in the most contaminated regions of the Great Lakes. Generally, along the St Lawrence these species are not experiencing noticeable health problems. However, there can be exceptions to this, and we try to relate the variations observed to those seen in the biochemical responses. The great blue heron and mudpuppy have now been selected as indicator species. In monitoring ecological systems, biological indicators are very helpful in understanding how wildlife species respond to the stresses of chemical contamination, as well as other factors such as habitat loss.

# Alien Invaders

## *The Trouble with Zebra Mussels and Other Exotic Species*

*Gregor Gilpin Beck*

*"Whether the establishment of aliens is accidental or intentional, the ultimate effect is a form of 'biological pollution.' All too frequently, these alien invaders experience explosive population growth, thanks to a lack of natural predators or competitors."*

STANDING ON SHORE and gazing across the lakes and rivers of the Great Lakes–St Lawrence watershed, it is difficult to fathom the complexity of ecological relationships existing below the surface of these waters. The aquatic communities that have co-evolved over millennia are not readily visible and are therefore poorly understood. The native plant, invertebrate, fish, and other species form an intricate web of life.

Exotic species can wreak havoc on the native flora and fauna. Whether the establishment of these aliens is accidental or intentional, the ultimate effect is a form of "biological pollution." It is true that many aliens fail to become established in new territories, and others may remain relatively uncommon. All too frequently, however, these "alien invaders" experience explosive population growth, thanks to a lack of natural predators or competitors.

The ecological impact of invasive exotics varies. Some compete with native wildlife for a specific type of habitat, such as starlings and house sparrows which aggressively exclude native birds from nesting sites. Other species completely change the nature of a specific habitats. For example, purple

Billions of zebra mussel shells line the beaches of our lakes and rivers. Below the surface, these exotic species are dwarfing populations of native clams and mussels, and causing serious environmental and economic problems.

Facing page: Zebra mussel shells. Photo: Gregor G. Beck

loosestrife, a deceptively pretty flower, quickly spreads through wetlands, choking out native plants and displacing wildlife. In rivers and lakes, exotic species have had staggering effects. Sea lampreys, a parasitic eel-like fish, entered the Great Lakes through the canals of the Seaway. Attaching to large fish species with a rasping, suction-cup mouth, lampreys literally suck the life-blood from their unfortunate hosts.

The effects of other exotic aquatics are less dramatic and horrific but no less damaging. Throughout the watershed, many non-native varieties of game fishes, including salmon and trout, have been introduced for sport; unfortunately, these species disrupt established food webs. Effects are wide reaching and may contribute to declining commercial fisheries. Changes in abundance of non-native predators also affect the numbers of smaller forage fish, such as perch and smelt. In some water bodies, such as Lake Ontario, virtually the entire fish community is artificial. The carp is another large exotic fish now widespread throughout the Great Lakes–St Lawrence watershed. Among other ecological concerns is the carp's ability to jeopardize wetland restoration projects by consuming and uprooting vast quantities of newly established aquatic plants.

Zebra mussels serve as one of the best examples of the complex and unpredictable impacts which exotic species can have on aquatic ecosystems. These coin-sized invaders were first detected in Lake St Clair in 1988 and within a few years had spread widely throughout the watershed. They are now found in lakes and rivers far removed from the Great Lakes and St Lawrence River proper, and have spread all the way down the mighty Mississippi – and beyond. Native to eastern Europe, the zebra mussel was transported to North America in the ballast water of ships. When this water was dumped prior to putting on cargo, the mussel got its foothold. Many other aquatic alien species have also become established this way; about one-third of all known Great Lakes invasions have occurred since the opening of the St Lawrence Seaway in 1959.

The zebra mussel alternates between a minuscule, free-floating larva and a hard-shelled, clam-like adult which attaches in large clusters to any solid surface. The larvae can remain free-floating for one to seven weeks, thereby increasing the potential for spread to new regions. In rivers as well as lakes, water currents can spread larval mussels hundreds of kilometres before the adult form settles and attaches to the bottom. Meanwhile, commercial ships and pleasure boats continue to travel the waterways, inadvertently spreading adult and larval zebra mussels. Since they can reach sexual maturity in one season and a single female can lay over one million eggs per year, the potential for population growth is extraordinary.

Most of the discussion regarding the impact of zebra mussels has centred on the considerable human and economic effects. Cottagers and vacationers are frustrated by mussel-covered rocks in swimming areas which cause cuts and reduce the enjoyment and quality of the waterfront environment. The economic cost to industry is just starting to be added up – and the final bill will certainly be in the hundreds of millions of dollars. Water intake pipes and other underwater structures quickly become covered with enormous densities of mussels. The water flow in these pipes concentrates both larval mussels and the suspended algae upon which they feed. Removal or abatement programs, including chlorine treatments, are being used to address the problem, although less ecologically harmful alternatives, such as heat treatment, should also be pursued.

More significant, however, are the staggering and less frequently reported ecological implications of this alien invertebrate. Large densities of zebra mussels can deplete dissolved oxygen from small water systems. In addition, zebra mussels feed voraciously on microscopic plankton at the base of the food chain, and therefore compete with native invertebrates and plankton-eating fish species for food. While this can increase water clarity, there are detrimental effects: fewer phytoplankton and zooplankton, therefore fewer small fish, and ultimately fewer predators, such as lake trout. In effect, zebra mussels are changing the flow of nutrients, and ironically, pollutants such as PCBs, in aquatic ecosystems. More and more nutrients are being diverted from planktonic to benthic food chains. This actually helps increase the abundance of some native bottom-dwelling invertebrates, which may in turn benefit bottom-feeding fish,

including carp, bass, and yellow perch. In some locations, bass and muskellunge may also be benefiting from more abundant underwater plant beds. Nonetheless, the overall fertility, or productivity, of the water is reduced.

Because of their tremendous abundance and ability to attach to a variety of surfaces, including the shells of native mussels and clams, zebra mussels are dramatically reducing aquatic biodiversity. Through a combination of fouling and competition for food, this alien has already virtually eliminated native, bottom-dwelling mussel species from Lake St Clair, Lake Erie, and most of the upper St Lawrence River. In the Mississippi watershed, about sixty of the 131 species of unionid-type mussels are now endangered or threatened with extinction. In the upper St Lawrence, the two most abundant mollusc species are both exotic: the zebra mussel and the faucet snail. Other exotic molluscs in the watershed include the Chinese mystery snail, European spaeriid clams, and also the quagga mussel (which dominates the benthic fauna in some regions).



The dense zebra mussel colonies on rocky underwater surfaces reduce populations of some snail and insect larvae, while enhancing others. Recently, zebra mussels have also been observed in areas with soft bottoms, which could help them to further expand their range – and their impact. On a more positive note, commercial ships are now required to exchange their freshwater ballast at sea, and pleasure boaters and anglers are being instructed on how to avoid accidental transport of mussels. In addition, some of our native wildlife appears to be acquiring a taste for these invertebrates. In highly infested parts of the lower Great Lakes, there have been significant increases in the numbers of diving ducks, notably scaup, scoter, and oldsquaw species. Other wildlife, such as map turtles, mink, various fishes, and crayfishes will feed on zebra mussels also, if preferred prey is not available. While these observations are encouraging, it is unlikely that natural predators will significantly reduce zebra mussels.

The cumulative effects of the zebra mussel are great. Unfortunately, it is only one of approximately 140 species of aquatic aliens now present in just the Great Lakes. Across North America, there are an estimated 4,500 non-native species of plants and animals, and many more are likely to become established in the future. As the ecological implications escalate further, researchers probe the impact of exotic species and continue to seek possible solutions and methods to predict and prevent future invasions.

Port of Montreal and St Lawrence Seaway. Photo: Phillip Norton

# The Montreal Archipelago

## *Biodiversity in Quebec's Urban and Industrial Core*

*Michel Letendre*

*"Rare are the large cities in North America like Montreal, Laval, and Longueuil where it is possible, just minutes from downtown, to observe such diverse and bountiful flora and fauna, often unique to Quebec ... This wonderfully diverse range is possible because of the local climate, the warmest in Quebec, and also the rich green waters of the St Lawrence River which join here with the brown waters of the Outaouais (or Ottawa) River."*

IMAGINE THE FOLLOWING SCENE, worthy of a first prize in photography. Against the background of a spectacular sunset, a duck is feeding with her brood at the edge of a marsh. A few metres away a muskrat nibbles on reed stems, and a great blue heron perches motionless, preparing to capture its prey. In contrast, the reflection on the water shows a very different picture in the background: skyscrapers and merchant ships in the Port of Montreal.

Rare are the large cities in North America like Montreal, Laval, and Longueuil where it is possible, just minutes from downtown, to observe such diverse and bountiful flora and fauna, often unique to Quebec. Here in Quebec's heartland, anglers can fish for trophy-sized muskellunge, walleye, bass, American shad, and brown and rainbow trout, and hunters can stalk migrant waterfowl. In winter, large numbers of residents enjoy ice-fishing, setting up veritable villages on the frozen Lac St-Louis. The yellow perch, which has an excellent flavour, is the species caught in greatest abundance.

This wonderfully diverse range of fauna and flora is possible because of the local climate, the warmest in Quebec, and also the rich green waters of the St Lawrence River which join here with the brown waters of the Outaouais (or Ottawa) River. The multiple tributaries of these waters, which include the Rivières des Mille Iles and des Prairies as well as the rapids at Ste Anne and Dorion, shape the many islands that create the Montreal Archipelago.

In total, ninety of Quebec's 116 freshwater fish species are present in the archipelago. Some of these species are prehistoric and fantastic – such as the bowfin. The male bowfin is often engulfed by a black cloud of offspring from which he darts ferociously to ward off predators. Another prehistoric fish is the longnose gar, its long pointed snout armed with dozens of sharp teeth. From the Atlantic, American shad swim up the St Lawrence and pass through the waters of the archipelago during the spring migration to spawning sites in the Outaouais River. A frenzy of anglers along the shores of the Rivière des Prairies turn out for this annual meeting with the great, shiny, silver-coated marathoner. Another long-distance traveller is the American eel, which feeds in the archipelago on its way back to the Great Lakes.

In addition to resident and migratory waterfowl, the great blue heron, the black-crowned night-heron, the double-crested cormorant, and the ring-billed gull can be seen from cars crossing the bridges of the archipelago. About three hundred species of birds can be seen throughout the year. Mount-Royal and the regional parks of the Montreal Urban Community, Longueuil, and the Rivière des Mille Iles are havens for both migratory and nesting birds.

Reptiles and amphibians are also abundant in this region. The common map turtle, rare in Quebec, lives in and around Lac des Deux Montagnes and represents one of the few viable turtle populations in the province. An equally significant site for reptiles is Ile Perrot, which provides important habitat for the rare northern water snake.

Montreal's archipelago has long been renowned for its wonderfully diverse flora, including large concentration of rare plants. In this region, 106 of the 183 priority species for the St Lawrence River can be found, eighty-five of which are at their northern limit. River-weed, for example, is found in the rapids of the Montreal Archipelago, and nowhere else in Quebec. This plant thickly carpets the rocks of the rapids, the only species of its family to inhabit the temperate boreal zone. The growth of plants is outstanding: the floodplains of the archipelago produce more biomass per acre than the best cornfield.

However, the large amount of human activity that has occurred throughout the archipelago since the first Europeans arrived has greatly altered natural habitats. By about 1970 it was estimated that this region had already lost 80 percent of its original wetlands. On the brighter side, the Archipelago of Sainte-Rose in the Rivière des Mille Iles represents one of the best preserved natural areas in the Montreal Archipelago. For this reason it has been aptly nicknamed the "Everglades of Quebec."

The creation of provincial and regional parks has protected some of the last remnants of highly valuable natural areas. The parks are visited year round by individuals who appreciate the rich beauty of the local fauna and flora. In return, many of them volunteer their efforts to ensure that these areas continue to be well preserved on the doorstep of a major urban centre.

Atlantic puffins, St Mary's Islands Seabird Sanctuary, Quebec. Photo: Gregor G. Beck

# Balancing Act

## *Ecotourism along the Shores of the St Lawrence*

*Dominique Brief and John Hull*

*"Ecotourism is booming, and regions throughout the Great Lakes and St Lawrence watershed are taking part ... The Kahnawake and Lower North Shore case studies illustrate how tourism is playing a small but important role in improving the health of the natural environment along the St Lawrence. These examples illustrate that through education, local economic incentives, and natural area protection, it is possible to find a balance."*

COMMUNITIES WORLDWIDE are promoting tourism to create employment, generate capital, and establish greater local control of activities in efforts to preserve natural and cultural resources. Ecotourism is booming, and regions throughout the Great Lakes and St Lawrence watershed are taking part. Communities on the Kanawake Native Reserve and the Lower North Shore of Quebec illustrate public and private efforts to build a healthier economy and environment through ecotourism.

The Kahnawake Native Reserve is located along the south shores of the St Lawrence, nestled between the island of Montreal and the town of Châteauguay. Recently efforts have been made within the community to address lack of employment, frustration with public image, and environmental issues. From these concerns came the idea of an ecotourism organization. Mohawk Trail Tours is a new business with a client base which is 90 percent European, 8 percent regional, and 2 percent from Asia, the Caribbean, and North Africa. The management of this group has a solid notion of how to contribute to the conservation of their land. This includes increasing environmental awareness for both native and non-native peoples, securing economic contributions from tourists, and channelling funds towards activities promoting the health of native culture and the land.

One of the topics discussed during walking tours is the environmental impact of past and future developments, including damages caused by the St Lawrence Seaway and the proposed plan to dam the Lachine Rapids. When the Seaway was constructed, farmland, clean water, and fish were lost. This had a negative impact on the diet and health of those who rely on these resources. If a dam is built on the Lachine Rapids, it is projected that this hydroelectric project would result in the flooding of large tracts of land within the Kahnawake reserve – land that is currently used for agricultural purposes. The goal of the public education project is to make people aware that major development has major environmental consequences which should be carefully evaluated *before* projects are begun.

Visitors are informed of the many local efforts to combat environmental problems. The Kahnawake Environmental Protection Committee's initiatives include providing free trees for planting, running a door-to-door recycling program, establishing the local community garden, and evaluating landfill waste. Mohawk Trail Tours also channels funds from ecotourism to the local cultural centre, through endeavours such as the sale of nature postcards. Management chose to contribute to this particular centre because its objective is the preservation and promotion of native history and culture, as well as the development of educational programs highlighting current environmental issues.

Mohawk Trail Tours has not overlooked the negative impacts that tourism can have on the environment. Tourists visit the old village and the more robust natural areas while the tour-guide describes the many historical, cultural, and natural highlights. In an effort to preserve ecosystem integrity, several sensitive natural areas are off-limits to both tourists and natives. These restrictions are based on ecological criteria such as the protection of critical nesting, breeding, and foraging habitats.

Far downstream in the northern Gulf of St Lawrence, the small, isolated communities of Quebec's Lower North Shore are experiencing serious hardship because of the collapse of groundfish stocks. In response, a small-scale ecotourism industry is developing around some of the oldest seabird sanctuaries in North America. The cooperative efforts of public and private organizations and the Quebec-Labrador Foundation (QLF) are helping to define a strategy that will provide employment and revenue while preserving seabird populations and the English, French, and First Nations cultures of this interesting area.

QLF has been working with over twenty local public and private organizations to produce visitors' guides in French and English celebrating the natural and cultural history of the region. The rugged landscape, the abundance of whales, Quebec's largest puffin colony, the site of the invention of the cod trap, and the crash site of the first transatlantic flight from Europe are just a few of the many attractions for visitors. The goal of the guides is not only to identify attractions but also to increase awareness of local services and businesses. Detailed community maps identify restaurants, accommodations, and information centres.

Income generated from the sales of guides is being used to help fund QLF's local conservation and development projects. The Marine Bird Conservation Project has been providing the youth of "the Coast" with programs on seabird biology, conservation, environmental education, and outdoor skills since the mid 1970s. The guides are also providing many new local businesses with an opportunity to increase revenues from tourism. Even though these revenues are small at present, marketing through the Internet, information centres, and word of mouth will help to expand the service sector of the local economy while ensuring that conservation objectives at the sanctuaries are met.

With the increasing number of visitors to the region, the guides also play a critical role in raising awareness of local wildlife and land-use regulations. These regulations are identified by the Canadian Wildlife Service and local residents as critical for safeguarding local flora and fauna. For example, permits must be purchased through local agencies to visit the sanctuaries or to pick cloudberries or "bake-apples," a prized local fruit used in jams and pies.

The Kahnawake and Lower North Shore case studies illustrate how tourism is playing a small but important role in improving the health of the natural environment along the St Lawrence. Ecotourism represents a new approach to integrating business and the environment. While positive examples of ecotourism are often few and far between, these case studies illustrate that through education, local economic incentives, and natural area protection, it *is* possible to find a balance.

# Blocking the Flow of Rivers

## *The Impact of Dams on Nature*

*Fred Whoriskey*

DAMS BRING ECONOMIC BENEFITS, but they also have high environmental costs. This dichotomy lies at the core of the disputes arising around the construction of virtually every new dam. Groups that value the services and economic opportunities which dams create will always be in conflict with other groups that value the services, and economic opportunities, and beauty provided by river systems in their natural state.

Dams collect water in reservoirs during times of plenty and make it available for use when precipitation is scarce. They are a boon for the construction industry, at least during their building phase. The water stored in the reservoirs behind dams can boost existing farm production and make farming possible in regions where it was once impossible. Impoundments also collect floodwater, minimizing the damage done to houses and roads. (On occasion, though, these systems can become stressed and overloaded, making flood conditions worse rather than better, as was the case with the Sageunay River flood of summer 1996.) And, of course, the water passed through turbines generates large amounts of electricity, which is essential for modern life. This power is relatively "clean" – no radioactive wastes, and none of the air pollutants that result from the burning of coal, oil, or wood.

However, dams dramatically change the river ecosystems on which they are located. These changes are for all intents permanent, because dams are seldom removed. As the reservoir floods, any vegetation that has not been cleared drowns. These plants and the organic matter in the inundated soils begin to decompose. This uses up the oxygen in the water, which can kill or drive away species that require high oxygen levels.

The flooding and decomposition trigger a big pulse of nutrients which fertilizes the reservoir and causes abnormally large phytoplankton or algae blooms. These blooms may be unwelcome for aesthetic reasons, and their respiration can strip more oxygen from the water. The flooding also releases elemental mercury from soils into the water. Under oxygen-poor (anoxic) conditions, bacteria transform it to the highly toxic methyl mercury, which is bioconcentrated as it passes up the food chain. While mercury toxicity from reservoir food chains is probably most familiar from studies of the Cree living near the La Grande complex in northern Quebec, it is a problem in all of our reservoirs including those that drain into the St Lawrence River watersheds. A prime example is the Manicouagan River reservoirs. People may not be able to eat large, predatory fish for many years until reservoirs purge themselves of the mercury.



As a reservoir fills, stream conditions disappear and the area becomes increasingly lake-like. Communities may be displaced from their homes, and land is permanently lost for other uses. Within the reservoir, stream fishes are replaced by species adapted to lakes. The water column stratifies in summer, with the deep water becoming cold and deoxygenated. When water is released from this cold layer to pass through turbines or irrigation channels, animals in the downstream areas can be cold-shocked. The dam also blocks the spawning or feeding movements of stream fishes, thereby eliminating them. The Manicouagan River dams, for example, blocked Atlantic salmon movements, and the wild, sea-run population was lost. Fish ladders or other passageways help mitigate problems, but some fish are still killed falling over the dam or passing through the turbines. If the fish must pass over or through more than one dam on the course of their movements, the chance that they will be killed obviously increases.

Even ocean ecosystems can be affected by dams. In Canada many of our hydroelectric facilities pass most of their water through the turbines in winter, when the need for energy for heating is greatest. In contrast, estuaries and coastal ecosystems are "used to" receiving big pulses of freshwater from the spring melt. These spring freshwater runoffs bring nutrients that help fuel coastal and estuary biological production at a critical time of the year. We are completely reversing the natural pattern of freshwater discharge.

This was a concern in the environmental hearings over Hydro Quebec's proposal to divert water from the Moisie River, on Quebec's North Shore, into the neighbouring Ste Marguerite River. The project also proposed adding a third generating station (SM3) to the two already present on the Ste Marguerite. A major issue raised during the public hearings was that we do not know what the long-term impacts of shifting peak discharges to winter would be, especially in regions where more than one river has been harnessed for power production. At present the SM3 station is being built, but a final decision on whether or not to divert water from the Moisie River has not been made.

Given the potential impacts of dams, and their potential benefits, a process for fairly and carefully evaluating the issue becomes crucial. Environmental impact assessments, as mandated by both federal and provincial laws, are increasingly filling these roles. The process can be intimidating, especially for local community groups. However, during the course of it, issues are aired, and projects are usually changed in some manner before approval, taking into account the concerns of the public. Whether or not the changes are enough depends on how much the project hurts you or benefits you. Some element of conflict will always remain.

Female common eiders with ducklings, St Lawrence estuary. Photo: Gregor G. Beck

# Great Rivers Meet

## *Stirring Life and Wonder in the Saguenay–St Lawrence Marine Park*

*Nadia Ménard*

*"Life abounds here, yet, as in other parts of the watershed, human activity exerts pressure in the form of pollution, coastal development, intensive navigation, fishing, and hunting. Seabirds, seals, and whales are the most conspicuous predators, and their great numbers are an indication of the exceptional conditions in this area. Microscopic plankton are at the base of the marine ecosystem. Beneath the dark waters, brightly coloured organisms and many species of fish depend on the plankton-laden currents for their food. Seabirds and waterfowl find refuge on the islands in the estuary and raise their young."*

THE ST LAWRENCE RIVER is a life-line reaching into the heart of a continent. Born from the Great Lakes and headwater regions, it changes as it flows from freshwater to the ocean. The estuary is that transition zone where freshwater gradually mixes with seawater in tidal rhythms, but here, at the mouth of the Saguenay River, the St Lawrence meets the sea with force. The upwelling of cold, nutrient-rich waters at the head of the deep Laurentian channel is responsible for the marine environment that sustains the plentiful life found in the area. The meeting of these water masses results in a dynamic system, continuously renewed and enriched by freshwater from upstream and by the upwelling of cold saltwater.

Life abounds here, yet, as in other parts of the Great Lakes–St Lawrence watershed, human activity exerts pressure in the form of pollution, coastal development, intensive navigation, fishing, and hunting. To enhance conservation efforts and to promote public awareness, efforts were initiated in 1990 to establish the Saguenay–St Lawrence marine park. The provincial and federal governments, in cooperation with local and regional organizations, worked together to achieve the collective vision of creating the park.

The Saguenay–St Lawrence marine park includes representative portions of the St Lawrence estuary and the Saguenay fiord. At the heart of the park, waters from the Saguenay and upper St Lawrence, as well as the upwelling currents, merge to sustain and concentrate an abundance of plankton and fish, which in turn attract much wildlife. Seabirds, seals, and whales are the most conspicuous predators, and their great numbers are an indication of the exceptional conditions in this area. Microscopic plankton are at the base of the marine ecosystem. Beneath the dark waters, brightly coloured organisms and many species of fish depend on the plankton-laden currents for their food. Seabirds and waterfowl find refuge on the islands in the estuary and raise their young. Marine mammals are able to store energy reserves vital for their survival. The ecological value of the area resides in its role as a critical feeding grounds, sustaining this great wealth and diversity of marine life.



The Saguenay fiord is an unique feature of the marine park. Formed by glaciers that ploughed their way down the continent during past ice ages, the high cliffs and deep sullen waters are permeated with mystery. A distinct freshwater surface layer supplied by the immense Lac St-Jean drainage basin upriver flows over cold saltwater that has spilled into the bottom of the fiord from the St Lawrence. Twice per day at high tide, waters from the estuary flow into the Saguenay, bringing nourishment and oxygen to these deeper waters. The ecological conditions in the fiord maintain distinctive communities, some of which include animals of Arctic affinity.

In this part of the Great Lakes–St Lawrence watershed the bounty of natural resources, the navigable waters, and the sheer beauty of the landscape has promoted human activity for centuries. The confluence of the St Lawrence and Saguenay rivers was a central meeting place and a cultural crossroads. Here, native peoples came to hunt marine resources and to trade, first amongst themselves and later with Europeans. The St Lawrence estuary was also a hunting ground for sixteenth-century Basque whalers. In the nineteenth century, forest exploitation along the Saguenay opened the way to settlement and industrialization. Today the area supports a lucrative tourist industry largely based on whale watching.

The marine park encompasses a total 1138 square kilometres (440 square miles), its boundaries, unfortunately, based more on human interests than on ecological factors. Furthermore, the park is located at the confluence of two rivers which are also major industrial waterways, therefore allowing the inflow of contaminants and making the area vulnerable to environmental spills. Tourism has become a major industry, and heavy boat traffic poses potential threats to wildlife. The balance between educating the public through wildlife observation activities and disturbing the animals in their natural habitat is a delicate one, difficult to achieve. This gives rise to the question of to what extent the marine park can attain its conservation objectives when economic activities generate the livelihood of so many local residents.

Of course, applying protection measures exclusively to the marine park territory is not sufficient. The case of the St Lawrence beluga whale clearly illustrates that creating a marine park is just one important step towards saving this endangered population. The home range of the beluga extends well beyond the park boundaries, and a major threat to the animal is toxic contamination imported from hundreds of kilometres upstream. The park demonstrates that the effectiveness of conservation measures depends not only on what is being done locally but also on activities within the entire watershed.

As a privileged observer, my fascination and my attachment to this exceptional area increases as I learn more about it. Originally from the Saguenay area, I was amazed to discover the richness of the natural and cultural heritage associated with the region. I have come to believe that educating the public about the beauty and the fragility of aquatic and marine ecosystems is vital to gaining support for conservation measures. Industrial and development interests have a major influence on decision making, but a strong voice is also needed for the sake of ecosystem conservation. Impacts are too often insidious or take place over the long term, compromising environmental protection for potential profit.

The Saguenay–St Lawrence marine park could serve as an example that ecosystem conservation can be attained alongside harmonious human activity. However, much

work remains to be done to attain these objectives. Through its presence, and through education and positive pressure, the true effectiveness of the marine park will be reached when the ecological benefits of the protected area have radiated far beyond its actual boundaries.

Facing page: Fishing dory near Matane, Quebec. Photo: Frozenrope

*Overleaf:* The North shore, Lake Superior. Photo: Frozenrope

# Watershed Perspectives

Photo: Frozenrope

# Bi-National Citizen Action

## *Grassroots Environmentalism in the Watershed*

*John Jackson*

*"Citizens will continue to be the driving force for the protection of the Great Lakes and St Lawrence watershed. Those living around the region are the ones most directly affected, and who enjoy and value the lakes, rivers, and surrounding lands for their multitude of essential and delightful facets. We have learned that it is only by bringing pressure to bear in a united and simultaneous way on all responsible jurisdictions and polluters that there is hope for protecting the watershed."*

CITIZENS' GROUPS have long been the driving force for the protection and clean-up of the Great Lakes and St Lawrence River. In every community across this vast and varied watershed, citizens are the watchdogs who first see the subtle and not-so-subtle changes in the environment. And they are the ones who launch the battles to force their fellow citizens, governments, and polluters to act more responsibly.

Over the past decade these people, who work primarily through volunteer grassroots groups, have joined forces basin-wide. They ignore political boundaries, and thus force governments to try to make a stronger effort to cooperate with each other.

As an example, citizens have become a powerful force for the implementation of the fundamental concepts in the Great Lakes Water Quality Agreement, first signed by the Canadian and U.S. governments in 1972. The principles of the agreement have become the rallying cries for the citizenry: "Zero Discharge Now!" and "No Time to Waste!"

Citizens worked jointly and effectively to ensure that the Great Lakes Water Quality Agreement would not be destroyed when the governments renegotiated the agreement in 1987. Great Lakes United, a coalition of 180 citizen, environmental, conservation, and labour organizations from the United States, Canada, and the First Nations, held nineteen citizen hearings around the lakes, creating a powerful voice and momentum that the governments could not ignore. For the first time citizens were directly involved in the negotiation of the agreement, with five citizen representatives sitting at the table when the final negotiations between Canada and the U.S. were conducted.

Citizens working together across the border have also played the lead role in increasing the profile and power of the recommendations of the International Joint Commission. The IJC's biennial meetings, where the state of the Great Lakes is assessed and where the agenda is set for the coming years, were once small gatherings of scientists and government bureaucrats. By the time of the Windsor biennial in 1993, the number of attendees had expanded to include almost two thousand, with the largest contingent the citizen activists.

Citizen involvement in this bi-national forum has dramatically increased the attention paid to the Great Lakes Water Quality Agreement and the IJC. Citizen frustration with the governments for failing to live up to their pledges in the agreement rose to a boil at the IJC's biennial meeting in Hamilton in 1989. Throughout the meeting, and especially during the eighteen hours of citizen testimony, the governments were condemned for failing to live up to their promises, and the IJC was criticized for failing to be a strong advocate for the Great Lakes.

Concerns and proposals from citizens have led the IJC to write biennial reports that are now less diplomatic and more direct in their style, and more passionate and determined in their proposals for strong, truly long-term solutions.

Working together has also allowed us to learn from each others' successes and failures and to develop common goals and strategies. These citizen goals and strategies are now the impetus behind many environmental programs.

Citizens will continue to be the driving force for the protection of the Great Lakes and St Lawrence watershed. Those living around the region are the ones most directly affected, and who enjoy and value the lakes, rivers, and surrounding lands for their multitude of essential and delightful facets. We have learned that it is only by bringing pressure to bear in a united and simultaneous way on all responsible jurisdictions and polluters that there is hope for protecting the watershed. The divisions that separate us are overcome by our mutual love of a shared treasure.

Photo: Deborah Freeman

# Sustainable Development
## *Balancing Environment and Economy*

*Michael Keating*

*"If we are going to learn to live more sustainably, we need to develop a greater understanding of how ecosystems work. We also have to learn which forms of consumption and waste disposal have the lowest impacts on the environment. This can be as simple as understanding that toxic chemicals poured down the drain cannot be treated by sewage treatment plants ... It can be as complex as choosing foods, clothing, and ways of travelling that lower our environmental impact."*

FOR MOST OF OUR HISTORY we humans had little impact on our environment. That changed with the dawn of agriculture about ten thousand years ago, which started the clearing of land and the building of cities and civilizations. Since the industrial revolution and the human population explosion began three hundred years ago, we have changed the environment on a regional, even global scale. In recent decades a growing number of people realized that pollution and resource depletion were hitting unacceptable limits. They saw that most of these problems were caused by the normal functioning of many of our businesses and by the type of consumption that was now part of the lifestyle of the industrial world. By 1987 the World Commission on Environment and Development was calling for rapid evolution into ways of living and doing business that are environmentally, economically, and socially sustainable. To do anything less is to keep undermining the environmental pillars that support us and to cheat future generations of the chance to live as well as we have.

What we need is development without the destruction of nature. This means using parts of the environment less intensively, so they will not become dangerously polluted or lose their ability to function normally. It means tailoring business practices and lifestyles to fit the ecological realities of our planet. We need fish but must stop over-fishing. We need paper and lumber but cannot keep cutting trees faster than they grow. We need transportation, but we also need land that is not paved, and air that is not fouled with exhaust gases.

To achieve sustainability, we must maintain the biological diversity that creates a healthy, productive, and stable environment and maintain such life-supporting ecological systems as the ozone layer, fertile soils, and fresh water supplies. We must maintain a healthy stock of renewable resources such as fish and trees, and we must use non-renewable resources in a way that does not deplete them for us or future generations.



Some forms of consumption, such as drinking water, eating, and housing ourselves, cannot be reduced below a bare minimum for human survival. Beyond that, our environmental impacts are greater or smaller depending on the technologies we choose, and the amount of resources we decide to consume. For example, a small car takes fewer resources to make and consumes perhaps half the fuel of a large car. Electronic communication can eliminate the need for some car trips altogether.

To live sustainably, we need to reduce the ecological impacts of a wide range of activities, from the way we grow food and turn trees into houses, books, and cereal boxes to how we design our cities. This means more clean industry, which will mean more efficient industry. It will require changes in the way we produce energy and run a wide range of businesses ranging from farming, fishing, and forestry to smelting, and petroleum and chemical manufacture. Industries can install more closed-loop processes that virtually eliminate the release of hazardous wastes. They can switch to less hazardous products, and use less toxic material. At home we also need to choose techniques, technologies, and products that have a lower total impact on our environment. We can conserve water and energy, reduce our use of toxic chemicals, and buy fewer throw-away products.

Sustainability is not a fixed state, because the world is not static. There are more people than ever before, and most want more of everything. The world's population grows by more than ninety million a year. Every three years the equivalent of another United States is looking to the earth's resources for food, housing, clothing, and an increasing number of consumer goods from toys to cars and from computers to jet planes. Earth's population reached six billion in 1999, and it may reach eight billion by 2020. It will likely keep rising – possibly to ten billion people by the time a child born in the late 1990s reaches retirement age. Since 1900 the global population has quadrupled, but fossil fuel use has increased by a factor of thirty, and industrial production by a factor of fifty. Most of this consumption took place in about thirty developed countries, but now it is being copied in more and more of the world's 150 other nations.



The Great Lakes–St Lawrence basin is a prime example of growth in developed nations. It is a population magnet, and since 1800 the number of people living in the region has increased tenfold. Its residents are among the world's highest consumers of energy and materials, owning more than sixteen million motor vehicles.

If we are going to learn to live more sustainably, we need to develop a greater understanding of how ecosystems work. We also have to learn which forms of consumption and waste disposal have the lowest impacts on the environment. This can be as simple as understanding that toxic chemicals poured down the drain cannot be treated by sewage treatment plants and keep going out into the environment. It can be as complex as choosing foods, clothing, and ways of travelling that lower our environmental impact. Eventually, we have to weave all these changes together into a pattern that will reduce our overall environmental impact to fit within the carrying capacity of the planet.

We can ask ourselves a number of questions to make choices for sustainability:

Is an activity sustainable in that it can be carried on indefinitely without running down its resource base?

Does it make economic sense while respecting the environment, or does it require the discharge of harmful substances or over-harvesting of resources to make money?

Is it socially and culturally sound? Is it being developed and carried out with the participation of those who will be affected?

Will this project or program use up non-renewable resources such as fossil fuels, and will it do so at a greater rate than some other project?

Will it use renewable resources at a rate greater than natural replacement?

Finding sustainable forms of development is one of the greatest challenges facing our society. We have begun creating organizations to help develop an understanding of the challenge and building partnerships and expertise. Numerous regional and local organizations are working for sustainable development. National organizations include the President's Council on Sustainable Development in the United States, and the National Round Table on the Environment and the Economy in Canada. At the global level there are such groups as the International Institute for Sustainable Development, Centre for Our Common Future, and the UN Commission on Sustainable Development.

Changing to sustainable activities requires time, imagination, and money. Often, the payoff comes partly in the form of energy and materials that are conserved, thus saving money. The benefits always include greater environmental security. Only with sustainable forms of living can we can guarantee ourselves such basic needs as fresh water, clean air, good food, shelter, sanitation, health care, energy, and, jobs into the decades and centuries ahead.

# Frog Reflections

## *The Ecosystem Approach to Conservation*

*Anne C. Bell*

*"The ecosystem approach will not enable us to avoid the difficult choices that lie ahead. Regardless of our approach to conservation, the survival of countless species, like the northern leopard frog, will have to be weighed against the fact that global warming, acid rain, persistent toxic substances, and the like form the very essence of our modern existence. Our reluctance to face up to required sacrifices may one day mean that the coming of spring will no longer be blessed with the low croaks and rumbles of leopard frog love songs. And to my way of thinking, that would be simply too high a price to pay."*

THE PLIGHT OF ANY LIVING CREATURE IS UPSETTING, but somehow the misfortunes that befall one's close neighbours seem that much more so. One of the more disturbing conservation issues of recent years has been the decline of northern leopard frogs, those cold-blooded, infinitely gentle familiars of my childhood. Although I witnessed the occasional martyrdom of many individuals over the years – blown up with firecrackers, impaled on fish hooks, butchered in biology class – only recently have I needed to worry about the very survival of their kind.

The northern leopard frog is one of many frog species worldwide whose numbers are plummeting. No one is entirely sure of the reasons for this disaster; however, acid rain, pollution, ozone depletion, global warming, and habitat destruction are all suspect. It could be that such stresses in combination are proving overwhelming for animals whose sensitive skin and amphibious habits render them particularly vulnerable. Regardless of the uncertainties, one thing is clear: frogs are in trouble because of the ways we, in modern industrial societies, live our lives.

In terms of conservation, the frog dilemma defies narrow

definition, ready explanation, and simple remedies. Thus, it reveals the inadequacy of traditional approaches that have focused on a species or a place in isolation from broader social, political, and ecological considerations. The issue demands instead that nature advocates recognize and address the staggering complexity of relationships and interactions that occur among humans and other species, and the worlds we inhabit.

One strategy which is gaining increasing support in conservation circles is known as the "ecosystem approach." It is based on the understanding that humans are a part of, not separate from, nature, and that our interests coincide with the health and welfare of non-human beings. As an example, the ecosystem approach has been a guiding management principle ever since it was included in the Great Lakes Water Quality Agreement in 1978. A response to the realization that no single factor was responsible for ecological problems in the region, its inclusion in this bi-national document has led to research and conservation initiatives that cross strict political, geographic, and academic boundaries.

As a concept, the ecosystem approach is still evolving, and for the moment it is interpreted in ways that both uphold and challenge the status quo. On one hand it is widely equated with more efficient control and manipulation of limited "resources" – or, in other words, maintaining the usefulness of nature for humans. On the other, it is hoped by some to imply a radical critique of humanity's proprietary claims to the rest of nature. Consumerism, profit maximization, and the human-centred worldview of mainstream society are held to be incompatible with the well-being of the natural communities of which we are a part. Between the two extremes, of course, lie intermediate points of view.

These varying interpretations are not surprising, given the broad range of interest groups that support the ecosystem approach. Included are representatives from government and industry as well as academics, environmentalists, and others. While such widespread backing of the ecosystem approach may encourage reform, from a conservation perspective it also represents a risk. In the long run it may mean that our efforts to protect non-human life will be subordinated to human concerns. So far, for example, with regard to the production and use of persistent toxic substances, the primary focus has been on harm to humans rather than to wildlife or natural processes. Further, in terms of research, economic interests have largely been dictating funding allocations and, thus, the ways in which the ecosystem approach is implemented.

The ecosystem approach bears great promise, but not as a vehicle for vague inclusiveness. Rather, its proponents should speak clearly about their values, desires, and intentions. For me the ecosystem approach means working from and towards an attitude of respect and caring for wildlife and wild places, regardless of their perceived utility. Humility with regard to human knowledge, abilities, and actions is also an important ingredient: when in doubt, we would be wise to err on the side of caution.

The ecosystem approach will not enable us to avoid the difficult choices that lie ahead. Regardless of our approach to conservation, the survival of countless species, like the northern leopard frog, will have to be weighed against the fact that global warming, acid rain, persistent toxic substances, and the like form the very essence of our modern existence. Our reluctance to face up to required sacrifices may one day mean that the coming of spring will no longer be blessed with the low croaks and rumbles of leopard frog love songs. And to my way of thinking, that would be simply too high a price to pay.

# Lessons from a Woodpecker

## *Parks and Protected Places*

*Jerry Valen DeMarco*

*"Any remaining opportunities to protect substantially natural areas must be seized immediately before the engine of development swallows them up. The various parks agencies must recognize that the absence of 'pristine' nature is no excuse to close the book on park creation. Areas around existing parks should be acquired whenever possible and new sites identified for protection and restoration."*

For many, a childhood in the country creates a lifelong concern for nature and natural landscapes. Perhaps it is strange therefore that as a dedicated nature advocate, I grew up in the shadow of industry – in Windsor, across the river from Detroit. We on the Canadian side were without many of the social and economic problems facing urban communities in and around Detroit, but our natural communities were and still are in a comparable position. Southwestern Ontario's natural habitat has been razed perhaps more severely than any other in Canada.

For those of us drawn to the natural world there were few opportunities to experience it aside from the fragments remaining in backyard gardens and roadside ditches. When I was four years old, however, a wayward Lewis woodpecker visited a telephone pole in our neighbourhood and unwittingly fostered in me a sense of wonder for nature that has never subsided. Nearby, world-renowned Point Pelee National Park provided a seemingly endless supply of newcomers for my budding obsession with bird-watching. It was fortunate that others in my family were also lured by the park's wildlife and beaches, for this made it easy to justify frequent visits. Those pilgrimages seem strange to me only now when I realize that

Facing page: Photo: Gregor G. Beck

driving an automobile through urban, industrial, and agricultural deserts to get to a park is not the ideal nature experience. Increasingly and unfortunately it is what we must endure.

Islands of green like Pelee underline both the importance and vulnerability of protected areas in the Great Lakes–St Lawrence watershed. In one sense they are of great value as sanctuaries for nature in a radically human-altered landscape. Viewed from the air, these parks must look like the only remotely hospitable stopovers available to travelling birds or monarch butterflies. At the same time, however, they are totally inadequate to ensure the long-term survival of natural communities. While the forces of nature continue to alter the shape of Pelee's sandy southern tip, its ecological heart is threatened by exotic species, pollution, high visitor use, and its isolation from other natural areas. Nearly one-third of Pelee's native reptiles and amphibians are now gone, and 40 percent of its flora is not native. The tiny fifteen square kilometre (six square mile) refuge is unable to do much more than provide a welcome break for weary migrants, bits of habitat for some species, and a scarce natural "commodity" for half a million human visitors a year. And Pelee's predicament is but one example of the destructive pressures on natural areas throughout the watershed.

It is true that governments and citizens on both sides of the border can take some pride and comfort in their parks such as Isle Royale, Pukaskwa, and La Mauricie, but if the full complement of natural diversity is to be retained, much more must be done. In Canada, the Endangered Spaces campaign championed by World Wildlife Fund Canada (WWF) and the Canadian Parks and Wilderness Society (CPAWS) envisions a protected areas system that includes representative natural areas for each of the country's distinctive regions. The campaign ambitiously aimed to complete such a network by the year 2000 with the support of provincial and federal governments.

In Ontario, the Wildlands League chapter of CPAWS has drawn up a protected areas blueprint for the achievement of the Endangered Spaces objectives in the province. Already, detailed mapping work for the northern reaches of the watershed – done in conjunction with WWF and the Federation of Ontario Naturalists – helped convince the provincial government to double the amount of protected areas in the region in early 1999. Their work in identifying the remaining gaps in ecological protection should be used to guide future protected areas planning.

In Quebec, WWF's Nathalie Zinger is endeavouring to spur on that province's modest parks and ecological reserves network through the development of a co-ordinated protected areas planning system. In addition to the need for new protected areas, the "green" myth of Quebec's wildlife reserves needs to be dispelled. These reserves are portrayed as large green blocks on road maps but actually still permit logging. This state of affairs, which creates a false sense of security, holds true in other so-called protected zones in the Great Lakes–St Lawrence watershed, such as state and national forests and the Algonquin and Adirondack parks.



On another front, the Nature Conservancy's Chicago-based Great Lakes Program is aiming to ensure that the protection effort throughout the Great Lakes region proceeds according to ecological priorities. This initiative has identified areas of natural significance in the watershed – some protected, most not. It prioritizes natural features, ranging from insect and plant species to complex ecological communities such as marshes, according to how threatened they are. An alarming number of features fall within the rare, imperilled, and critically imperilled classifications and point to the sense of urgency facing ecological protection in the basin.

An even more challenging task is being undertaken by the proponents of the U.S.-based Wildlands Project. While similar to the Endangered Spaces campaign in many respects, its time horizon is much longer. The protected system envisioned is one that will not only keep all the parts but also reassemble them according to what once existed. It provides the long-term vision that complements the shorter-term campaigns.

In many parts of the watershed, the option of protecting significant wild areas has already been substantially foreclosed by the historical transformation of the landscape into a patchwork of human uses. Here, the strategy must be to restore nature by weaving networks of protected remnants and restored sites. And those activities must be further enhanced by environmental initiatives focusing on problems such as pollution, resource over-exploitation, and inappropriate land use.

Susan Crispin of the Nature Conservancy recognizes the inextricable link between protected areas and the larger tasks of improving environmental health and restoring damaged areas. She points out that "the good work that is underway to control toxic discharges, restore fisheries, and remediate Areas of Concern can be complemented by identifying relatively intact systems that support key biological components, and undertaking local action to sustain them." This diverse approach may very well be the only way we can consolidate the gains that have been made.

The first step must be for governments to make nature protection rather than recreational use or tourist revenue the priority for the various parks systems in the watershed. Any remaining opportunities to protect substantially natural areas must be seized immediately before the engine of development swallows them up. The various parks agencies must recognize that the absence of "pristine" nature is no excuse to close the book on park creation. Areas around existing parks should be acquired whenever possible and new sites identified for protection and restoration.

Innovative land-use planning processes can also be used. A bold step was taken in Ontario with the creation of the Niagara Escarpment Commission whose mandate is to base land-use planning on environmental protection instead of economic development. Similar "nature-first" approaches need to be applied elsewhere in the developed landscape. Regrettably, recent government cuts have seriously limited the commission's effectiveness.

The nearly forgotten marine parks agenda must be rejuvenated as well. Progress has been limited to Lake Huron and Georgian Bay's Fathom Five and Quebec's Saguenay–St Lawrence marine parks, along with public consultations on a protected area for western Lake Superior. The establishment of marine protected areas provides a wonderful opportunity for the two Great Lakes nations to cooperate and recognize the international importance of the boundary waters. Considering our nations' long history of good relations, it is astounding that we have failed to either protect the basin's water quality or to provide marine protected zones. The easy route of establishing reserves that continue to permit commercial fishing and other major industrial practices, however, must be resisted. Rather, so long as the watershed's lakes and rivers are ravaged by overuse, pollution, and over-fishing, marine reserves offering real protections can play an important conservation role for nature.

The key element in effecting substantial change is citizen involvement. Grassroots support must stimulate a new era in protection ranging from providing backyard habitat for wildlife to supporting large-scale protection and restoration initiatives. Unlike less densely populated regions, complex landholdings in the Great Lakes–St Lawrence watershed make park creation much more difficult and necessitate local involvement and support. As communities take action, natural networks and corridors can be created to foster ecological protection. A sense of home place must be recreated – not only ecologically but also spiritually and emotionally – so that humans again realize that they are part of a complex and wonderful natural community.

It is clear that we still have a long way to go before we can truly proclaim our societies to be environmentally sustainable. Restored natural landscapes and human connections to them have to be the ultimate goals, but how can we get there? In the immediate future, a major challenge will be to ensure that in our search we do not trample nature underfoot. That is, we must ensure that nature is not harmed while we endlessly deliberate about how and when we will change our ways.

The parks systems will be one means of helping to ensure that, whichever paths we choose, the natural communities native to the Great Lakes–St Lawrence watershed will survive. Protecting and expanding parks will help show that we are truly on the road to ecological sustainability – and sanity. If we do not act now, then not only will we fail our children by preventing them from developing their sense of wonder for nature but we will also fail nature itself. That is an ending I do not want to see for the story of nature that started, at least for me, with a wayward woodpecker searching for its natural home.

Photo: Gregor G. Beck

# Learning the System from Water

## *Life Lessons of Environmental Education*

*Kevin J. Coyle*

*"Learning our waters – where they are, how we use them and what we do to them – teaches us important lessons. These should be among the basic lessons of life. As we become more aware of our water sources, we also become more able to keep them clean for those future generations we so carefully tuck into bed each night. Without this connection, people are less likely to take actions to keep our waters clean. Education about the environment – in the classroom and in a lifelong setting – is vital to understanding our own place in the world. And water holds the central place in that education."*

IN MANY HOMES, putting the children to bed is a nightly ritual. And, most often as not, the little ones want to be carried. But how many parents ever stop, mid-lug, to think about exactly *what* they are carrying? Indeed, our young ones (and ourselves) are fully 70 percent water. For many this means they are made up largely from a local river or lake. To make this point, I sometimes introduce my own children or grandchildren by naming the rivers that occupy their bodies – the Ohio, the Delaware, the Potomac. This illustrates the simple point that we are more directly connected to our waterways than we may think.

Learning our waters – where they are, how we use them and what we do to them – teaches us important lessons. These should be among the basic lessons of life. As we become more aware of our water sources, we also become more able to keep them clean for those future generations we so carefully tuck into bed each night. Without this connection, people are less likely to take actions to keep our waters clean. Education about the environment – in the classroom and in a lifelong setting – is vital to understanding our own place in the world. And water holds the central place in that education.

Public opinion research conducted from 1992 through 1997 for the National Environmental Education and Training Foundation reveals some startling facts about Americans' knowledge of the environment. This research shows that while people care more about the quality of their water than any other environmental concern, they know little about it. Between our caring and our actual knowledge is an almost unbelievable gap. For example, only 54 percent of adult Americans know for sure that the oceans are *not* a major source of fresh water. This underscores how few people are clear about where their water comes from – and what is in it. Awareness and knowledge about water resources are probably similar in Canada.



The research also shows that, despite the fact that the main cause of pollution in the United States comes from rain and snow washing pollution from the land (roads, farms, lawns, and lots) into rivers and lakes, the average person still thinks of industry and cities as the main problems. It is easier to blame a local factory for a pollution problem than it is to look critically at our own daily activities and those of our neighbours and community leaders. So, if the everyday actions of people are the main problem with water, then public education must be a major part of the solution.

Imagine a future in which environmentally literate people are conservative in their use of water and careful about what goes down their drains and on their lawns. Imagine an educated public that practises these measures at home, at work, and at school. In classrooms across the nation, learning about water is a cornerstone of environmental education. Programs such as the Council for Environmental Education's Project WET and Give Water a Hand at the University of Wisconsin give teachers supplemental training on how to engage students in short courses on the importance of water.

Getting the public to pay attention to water and environmental issues requires more than training programs and simple instruction. It even goes beyond telling them that the health of the planet depends on it. The public can perhaps be best persuaded by illustrating how environmental education benefits everyone directly. This idea was demonstrated by public opinion research conducted for the conservation organization American Rivers in 1994. The survey showed that roughly one-third of adult Americans cared deeply about preserving nature – wetlands, wild areas, streamside habitats – but that nearly nine out of ten were *deeply* concerned about the quality of water. The depth of caring about the environment is encouraging, but the degree of concern for water is even more impressive.

When Americans were asked whether protecting the country's freshwater drinking supply should be a national priority, nearly 95 percent said yes. This is good news for those of us concerned with the conservation of rivers and lakes in North America, because our continent contains about one-half of all the fresh water in the world. Canada alone contains one-third of the globe's fresh water. The bottom line is that there is overwhelming support for the protection of our waters, especially once people make the critical connections. The reasons why environmental education is a must for all members of society are limitless.

Research conducted by the State Education and Environment Roundtable of San Diego, California, in 1996 demonstrates that environmental education makes children better learners. Students in some forty schools became much more enthusiastic and achieved better grades in math and science when they learned through the lens of the environment. This was corroborated in a separate study at an Illinois high school: students who learned to apply their science and math skills through in-depth study of a local pond ecosystem also had higher test scores than those who learned conventionally through the classroom only.

The Roundtable study also indicated that environmental education makes teachers more engaged and effective by encouraging team teaching and cross-over between several conventional disciplines. Fully 95 percent of the teachers surveyed showed considerably higher enthusiasm when they employed environment education in the classroom. Enthusiastic teachers mean better-educated kids and happier parents.

A key element of environmental education is that it encourages reasoned and responsible actions on the part of all individuals. There is great value for children, who live in a world of television and video and computer games, in learning what they can do personally to protect their local environment – plant a tree, save water, turn off the lights, and recycle. There is also a real value for parents in having their children learn to be more responsible and caring. Few activities in education give our children such an opportunity to look beyond themselves and become more generous.

Some critics think that children get too much negative information from environmental education and teachers. They say this gives kids a "doomsday" perspective. Studies at the University of Michigan and elsewhere, however, show that children who learn that their actions can make a real difference become more hopeful about the future. Good environmental learning balances the problems with solutions, and therefore creates hope and builds self-esteem. A child who learns that she or he can make a positive difference in the world through small actions ends up feeling more powerful.

Contrary to the beliefs of many, environmental education is not pitted against business and in fact can often be in support of business. Most long-term business decisions involve an environmental impact, and yet most business schools do not encourage environmental learning. This is starting to change. Some of the most prestigious graduate business schools in America, including Harvard, Yale, and Stanford, recognize that adequately preparing business leaders for a competitive future means teaching them about environmental impacts and technology. Business experts at companies such as AT&T, Lucent Technologies, 3M, and others go further by asserting that the businesses and business leaders who have a good handle on environmental technology will also be the most profitable and effective in the next century.

In 1993, the City of Milwaukee, Wisconsin, experienced a deadly outbreak of the water-borne parasite *Cryptosporidium* in its public drinking supply. More than 400,000 people became ill with flu-like symptoms, and 103 people died. The problem was more than the parasite and contamination from agricultural runoff. A very big part of the problem was that public health agencies took three weeks to determine its aquatic origins. The profound lack of environmental education in the health-care community resulted in the root cause being missed. The Centers for Disease Control note that environment-based outbreaks of disease could be more quickly identified if doctors and nurses had more than an average of four hours of environmental education in their many years of medical training.

People need to know more about the environmental risks of daily life. We all have a right to know what is in our water, our air, and the products we use, and what's the impact of toxic substances in our area. We have a right to more than awareness of these facts: we need to understand this information through good education.

In early 1997 massive floods in the American Great Plains and the Canadian Prairies forced tens of thousands of people from their homes. A major flood along the Mississippi River in 1993 had caused even more damage. Most communities in Canada and the United States are located along rivers, but few people are aware of the risks of building in the flood plain. The lack of environmental education about the processes of nature and the possible impact of inappropriate flood-plain development extends from average citizens to public officials. There is a definite need to better educate land developers, planning boards, town councils, and others on the proper management of flood-prone areas.

The need for greater awareness relates to every environmental and health concern conceivable. For the average individual, this knowledge and understanding bring benefits that include lower taxes and more attractive homes and communities. But such learning is ultimately more complicated.

We know that people care deeply about the quality of their water. Research and surveys clearly support this view. Unfortunately, people care much less about the protection of wetlands and other natural features. This is true even though it is well established that wetland systems including marshes and swamps play a vital role in protecting water quality because of their ability to absorb excess nutrients and toxic pollutants and reduce the risk of floods.

But to understand the connection between wetlands and water quality, one must learn a good deal about ecosystems and how they function. Ecosystems are those places of interlocking connections between animals and plants and soils and water. Because of the infinite interdependencies, these are complex and inherently difficult to understand. Ultimately, however, we must learn about at least the basics of ecosystems if we are to learn about our own connections to the natural world. If we care about water quality, then we need to better understand how wetlands and streamside vegetation clean water naturally. We need to know that healthy ecosystems are our best insurance policy for a healthy future. What we will undoubtedly learn as we become familiar with these systems is that we affect them and they affect us. We are in them and they are in us. By learning the "system" of our waters, we are ultimately learning about ourselves.

# Why Rivers? Why Watersheds?

## *Life-Lines of the Ecosystem*

*Peter Lavigne and Stephen Gates*

*Everybody has to go down the river some time. What river, well some river, some kind of river. Huck Finn said that and if he didn't say it, he should have said it. If he didn't, I will.*

Edward Abbey, *One Life at a Time, Please*, 1988

ED ABBEY HAD A WAY OF GETTING RIGHT TO THE POINT. When we cut to the chase about environmental protection, no matter what the issue, a river is always there. Like the veins and arteries connecting the life-giving functions of the heart to the human body, rivers comprise the ecological infrastructure of the continents. They are the roads and pipes of our natural systems, and the veins and arteries of the watershed body.

Rivers' life-transporting functions help to determine the health and ultimate survival of the entire watershed ecosystem. Rivers provide natural valley flood storage and wetlands and habitat for a wonderful diversity of aquatic and land species. Rivers supply tremendous natural vistas, impressive displays of nature's power, and conversely, contemplative opportunities central to our lives as thinking humans – attributes needing far more protection than they currently receive.

Rivers also perform a larger function in the natural realm. They connect the mountains to the sea, link headwater areas to lowlands, and provide a continent-wide system of pathways for the movement of plants and wildlife. This mixing, to the extent that it can occur on highly developed rivers, strengthens ecosystems by connecting natural areas – mountains, coasts, forests, and refuges – one to the other.

In many ways river systems also serve as the primary natural economic infrastructure: sources for waste disposal, power supply, transportation corridors, drinking water, and recreational use. Unfortunately, since 1600 the history of human development in North America and throughout the world shows an ever-increasing destructive capacity affecting natural riverine systems. The veins and arteries are denuded of their supportive organs while simultaneously becoming overloaded with sediment and other waste products.

Rivers, though, are part of complex watershed systems, with their own intrinsic value and ecological importance. They are also a dynamic and essential component of natural food chains. Despite the importance of watersheds in nature, the concept itself is poorly understood by the public. Most people have no idea what a "watershed" is: many equate watersheds with "water closets" or even fancy outhouses. Put simply, a watershed is the land from which water drains into a river, lake, or other water body. *All* land, therefore, is part of one watershed or another.



## Watershed Address and Ecological Literacy

We all know our mailing address, but how many people know their ecological address or their watershed address? These are tougher questions, and certainly more important ones from an environmental perspective.

What exactly is an ecological or watershed address? For our purposes, an ecological address may be defined as one's place in the watershed. Knowing your ecological address, your watershed address, means knowing your relationship with the waterways around you. Where does your drinking and washing water come from? What affects its quality and quantity? Where does it go when you are done using it or when you flush your toilet? What local sub-watershed do you live in, and to what major watershed do you belong?

Those of us who have lived in the Great Lakes–St Lawrence watershed have some idea of the wonders, vastness, and diversity of this region. But how many of us really know what it means to live in and be dependent upon this watershed?

We have heard a lot of talk about "cultural literacy" over the last decade from the likes of Allan Bloom and others. We rarely hear about, at least in broad public debate, ideas essential to physical survival in the new century. Ecological literacy, the knowledge of our ecological address and relationships, raises far more important issues than does cultural literacy (as defined by Bloom) for our ability to thrive in comfort over the millennium.[1]

Ecological literacy is fundamental to our work as citizen activists, environmental engineers, resource scientists, politicians, and regulatory administrators. It is also fundamental to the responsibility of every individual to make informed decisions about quality of life and community in a democracy. Do you know your ecological address? Do you know exactly where you live? Knowledge of our ecological address shows an understanding of our place in the ecosystem. Knowledge of our place in the ecosystem clearly indicates an understanding of the interconnectedness of the human and natural environment.

Consider, for instance, the ongoing public debates in the United States about the claimed unconstitutional "takings" of private property without compensation that supposedly result from passage and implementation of common, limited, and generally timid environmental protection measures. A basic level of ecological literacy throughout society would likely render this debate moot. Ecological literacy highlights the absurdity that individual property owners (you, me, and our neighbours) should be paid to protect the shared natural resources of land, air, and water which are required to sustain all life on the planet.

Canadian Otto Langer described the debate well in comments to a workshop on urban stream protection, restoration, and stewardship in the Pacific Northwest:

> The greatest weakness is that we do not apply the proper degree of protection to the watershed which gives rise to stream processes that create habitat. Instead, we apply remedial solutions such as stream restoration after the problem has developed ... We do this because society does not have the will to ensure that good land use planning will protect

key features of the watershed … If we do not take a very different approach in protecting our streams, they will continue to be victims of our growth dependent socio-economic system. The development rights of the landowner greatly outweigh the collective rights of the public to have healthy watersheds. However, it is the healthy watershed that is the true indicator of a sustainable future and a high quality of life.[2]

## Ecological Shock



> Ecologists perpetually talk about the interdependence of nature and lip service is given to this notion on Earth Day, but, in practice, environmental problems are approached one fragment at a time, not as a complex, multivariate, interdependent landscape. The coexistence of technology and biodiversity depends on switching from a fragmented to a landscape view.
>
> John Cairns Jr

Current trends reveal a crisis in biodiversity and an inability of current environmental protection laws and efforts to protect endangered species on either a piecemeal or systematic basis. Studies by the Nature Conservancy, the American Fisheries Society, and the National Research Council of the National Academy of Sciences illustrate that flowing freshwater systems sustain far more damage than terrestrial systems. Rivers are central to watersheds, so they suffer from environmental damage to land and water habitats – impacts that become amplified further downstream.

Nature Conservancy studies in 1990 and 1997 concluded that 37 percent of all native freshwater fish species in the United States are threatened or endangered, as are 65 percent of freshwater crayfish, and 67 percent of all freshwater mussels. Studies reported by the American Fisheries Society in 1991 and updated in 1997 have found equally disturbing facts for anadromous fish, concluding that 214 salmon and steelhead fish stocks in the Northwest are threatened, 101 of these near extinction. Some eastern species, including Atlantic Salmon and shad, are gone from most of their original spawning grounds.

The major causes of species decline include the loss of riverine habitat, dams, water diversions, river channelization, deforestation, development, and other habitat destruction. Because many ecologists believe that rivers are true indicators of general ecological and watershed health, these precipitous declines take on an even greater significance. Indeed, while at least ten native U.S. fish species went extinct between 1979 and 1989, no terrestrial species are known to have become extinct during that same period.

Similar concerns threaten Canadian rivers, although fewer studies and inventories have been completed. In southern regions including the Great Lakes–St Lawrence, urbanization is a major threat to rivers. Over-exploitation of commercial stocks and resource management remain serious issues in both inland and coastal fisheries. The Recreational Fisheries Institute of Canada reports hundreds of recreational fish stocks in jeopardy, with six species threatened and four in danger of extinction. The leading causes of fisheries decline are identified as human destruction of fish habitat, pollution, proliferation of non-native species, over-exploitation, and neglect. Generally, the ecosystems approach is preferable for preserving the diversity of fishes rather than focusing on one or two economically profitable species.

River systems across the United States are profoundly degraded. Human impacts have grossly altered their physical and chemical characteristics. A recent study by the National Research Council concluded: "Given that there are well over 2.5 million dams in the United States, only a small probability exists that a drop of water could make its way from its cloud of origin, over the land surface through the drainage system, and back into an ocean without passing through a man-made structure." The EPA reported in 1998 that 57 percent of U.S. watersheds do not meet at least some water quality standards and 40 percent of rivers, lakes, and estuaries do not meet the fishable/swimmable standards required by the Clean Water Act. The most recent available data also show that in 1994 more than 45 million people (20 percent of the U.S. population) were served by community drinking-water systems that violated health-based requirements at least once during the year.

With so much of the world's fresh-water resources, Canadians generally consider their water supplies very abundant. Some regions, however, are already beginning to stretch the limits of their water resources. Approximately 60 percent of Canada's fresh water drains north, while 90 percent of the population lives within 300 kilometres (180 miles) of the Canada–United States border. With some of the lowest water prices in the world, Canadians are now being required to re-examine this resource which has been taken for granted for far too long. Water conservation strategies include higher water rates, water meters, conservation technologies, and public education campaigns.



Degradation of river systems has become a serious concern, particularly in densely populated southern Canada. The main water problems in Canada are generally not related to quantity (although this issue is heating up also) but rather to issues of degraded water quality, disrupted flow regimes, and ecological balance. Urban streams and rivers have experienced significant habitat loss, as well as pollution problems resulting from sewer overflows, nutrient loading, and contaminants. The large area of impermeable surfaces in urban areas exacerbate the problem. Even northern rivers, which are commonly perceived as "pristine," are not immune to deterioration. In recent years, however, public awareness of the importance of riparian vegetation for streams has been steadily increasing, and many communities are establishing or rehabilitating "greenways," "blueways," and other wildlife habitats and corridors.

Additions to the "Ecosystem Collapse" file accumulate regularly. Included in the file is testimony by Dr James R. Karr, professor and former director of the Institute for Environmental Studies at the University of Washington. Dr Karr's 1992 testimony to the U.S. Congress, accompanied by a grim list of statistics, stated, "Simply put, the ecology of North American rivers has been decimated by the actions of human society. But river degradation goes beyond the loss of species. Sport and commercial fisheries of the U.S. have also been decimated by human actions during this century. Commercial fishery harvests in rivers such as the Columbia, Missouri, and Illinois, have declined by over 80 percent during this century."[3]

Emily Yoffe's stark cover story for the *New York Times Magazine*, entitled "The Silence of the Frogs" (13 December 1992), provided yet another illustration of the ecological shock affecting North American freshwater systems, namely the global collapse of amphibian populations. Studies collected by the Declining Amphibian Populations Task Force in Corvallis, Oregon, show almost one-third of North America's eighty-six species of frogs and toads appear endangered or extinct. In the article, David Wake, director of the Museum of Vertebrate Zoology at the University of California at Berkeley, says, "My theory is that it's general environmental degradation. That's the worst thing. Frogs are in essence a messenger. This is about biodiversity and disintegration, the destruction of our total environment." Follow-up reports published in *Conservation Biology* and elsewhere suggest that the decline in amphibian populations results in part from the continued deterioration of the ozone layer in Earth's atmosphere, along with a myriad of other causes including chemical contamination and loss of habitat. While these are very broadly based observations, they certainly apply to the Great Lakes–St Lawrence Watershed.

*Entering the Watershed*, a paradigm-shifting book by the Pacific Rivers Council, contains a comprehensive collection of ecological crises and policy recommendations for North American rivers. Documented impacts include the estimated disappearance of 70 to 90 percent of natural riparian vegetation due to human activities, and the fact that 70 percent of U.S. rivers and streams have been impaired by dams and other flow alteration.

Taken together, or even individually, scientific studies provide compelling evidence and a warning that the integrity of many of our natural systems, including rivers and their watersheds, may be overloaded and in some form of ecological shock.

## Traditional Approaches to River Protection

Wilderness and river conservation movements often converged in the 1960s and early 1970s with a focus on preventing dams in the few stretches of rivers not already dammed (or "damned"). Occasionally, we also focused on protecting land along the banks. Through the 1980s, traditional technical fixes accompanied these efforts, including sewage treatment plants, discharge permits, wetlands mitigation, and restoration.

Many of those traditional "fixes" merely put conditions on the new development or existing degradation to mitigate ongoing problems. Mitigation, however, does not address the root issues. It only slows the harmful effects addressed by discharge permits, dredge and fill permits, and development and habitat destruction. Meanwhile, pressures from population growth and development sprawl continually increase. These traditional approaches also involve enormous time and effort with provincial, state, and federal agencies which live and breathe technical fixes. In fact, if one takes a cynical view of environmental law protection, most of these laws were set up to implement discharge permits and enforce the degradation of our water, air, and land.

Professionals in agencies, national environmental groups, and local river groups become caught in this pattern over and over. We spend so much time fighting the battle of the day trying to reduce damage from developments or put a technical fix on the end of a sewer pipe that we miss the larger issues so clearly illustrated in scientific studies and in our rare walks along rivers.

## New Approaches

River-oriented, community-based stewardship began in Canada well over a half century ago, as public concern about river degradation increased. One of the first organizations was the Grand River Conservation Authority, which works in partnership with municipalities, citizens, and other government agencies in the watershed to solve flood, low flow, and water quality problems. After years of public pressure, provincial legislation formally established Conservation Authorities in 1946 to undertake programs for natural resource management, based on watersheds. There are now thirty-eight Conservation Authorities operating in watersheds in which 90 percent of the Ontario population resides. Their work includes reforestation and sustainable woodlot management, watershed strategies and management, flooding and erosion protection, acquistion and management of sensitive lands, and provision of education and recreation opportunities for all Ontario residents.

The Conservation Ontario watershed model has received worldwide recognition for its foresight and holistic approach. The watershed is now widely recognized as the best ecosystem unit for the management of natural resources, especially for water quality and pollution issues. In Ontario alone there are now over three hundred watershed management and strategic plans in place with member municipalities of Conservation Ontario.

In other parts of Canada, various forms of river-basin planning have been underway since the 1960s. For example, in Western Canada, the Fraser Basin Council and Partners for the Saskatchewan River Basin are multi-partnership groups with representation from First Nations, governments, business, and environmental groups. There are also hundreds or perhaps thousands of citizen-based watershed groups across the country.

In 1889 U.S. Army major and scientist-explorer John Wesley Powell proposed that western state boundaries be formed along the natural boundaries of river watersheds. Few people listened, and state and other political boundaries were formed according to a higher human "logic." Nearly three-score years later, citizen-based non-governmental river protection organizations in New England and other scattered areas began to heed Powell's advice. These groups have led a quiet revolution in environmental management since the late 1940s. Some have begun to implement educational programs stressing the fundamental interconnections between water quality, water supply, wetlands, air quality, and wildlife habitat. Watershed associations, including the Connecticut River Watershed Council, the Housatonic Valley Association, the Merrimack River Watershed Council, and many others, are at the front of this movement. These groups restore and protect the environment on an ecosystem basis, using river watersheds as the basic unit.

In the Great Lakes–St Lawrence River watershed, local and regional groups, including the Tip of the Mitt Watershed Council in Michigan, Rivers Unlimited in

Ohio, the Lake Michigan Federation, the River Alliance of Wisconsin, and the newly formed Minnesota Rivers Council, led a citizen-based renaissance in watershed protection during the 1990s. Governmental efforts on both sides of the international border are also joining in citizen and watershed based initiatives.

## Challenges in Watershed Protection

> Step back and look at this place ... Look carefully and you will see the myriad patterns of ridges and valleys etched with riverine paths pulsating to the rhythms of life; the cascading, trickling, gurgling, babbling, gushing, roaring, seeping, lapping, crashing, and silent rhythms of life.[4]



"Watershed" was one of the environmental buzz words of the 1990s and will no doubt lead environmental protection efforts in the new millennium. Debates on what watershed approaches mean politically and geographically echo through the halls of government, the offices of conservation organizations, and the pages of leading journals throughout the continent. The discussion is not, however, particularly new. A century ago, spurred by Major Powell's reports, political and environmental watershed boundaries were debated in the same wide range of venues including federal agencies, the U.S. Congress, and journals.

The time for watersheds – comprehensive, integrated environmental and political approaches to our river ecosystems, improved upon, refocused, energized, and revised – has arrived again, one hundred plus years later. A major challenge for the river watershed conservation movement includes gaining greater public understanding of the role that natural rivers and streams play in enhancing daily life. In particular, we need to communicate river watershed protection in ways with which people can relate.

We relate most easily to what we can taste, touch, and feel. Waste products and recycling are popular and easy to understand because we have to deal with them each and every day as part of living. The importance of river molluscs (clams and snails) and macro-invertebrates (bugs) and their relationship to a healthy and natural environment is a more difficult concept for the general public to grasp.

Effective education means going beyond the tangible daily issues of recycling and household wastes. It includes helping the public understand the importance of topics such as aquatic invertebrates, which are essential in food chains. The message must be clear that ensuring clean water and protecting rivers also means effectively addressing population growth, urban sprawl, air quality, solid waste disposal, and a myriad of other issues throughout the watershed. These topics must become a part of our culture and a part of our ecological literacy.

## The Question of Population Growth

Effective watershed protection involves a tough step-by-step process and the allocation of precious resources and limited staff. It means stepping back a little, trying to figure out the critical and global issues for the watershed. It also means making tough choices to cut back on time spent on day-to-day mitigation efforts, and in the state and provincial permitting statutes, and spending more time and effort on political change that enables comprehensive approaches to solving the broader issues. Watershed protection means educating adult decision-makers to regional issues and figuring out useful and innovative ways to adapt governmental boundaries to drainage basins and multiple jurisdictions. Most importantly, effective watershed protection and restoration mean focusing public attention on solutions for the most critical environmental issues.

We can spend lots of time and money in new environmental studies and initiatives determining the exact chain of events driving freshwater species into extinction, and in some instances perhaps we should. But it does not take a nuclear physicist, a rocket scientist, or a plant or fisheries ecologist to see the single most important factor in ecosystem degradation today. It is a factor which, if not addressed effectively and promptly, may quickly overwhelm every other action undertaken by humans to reduce environmental damage and restore healthy ecosystems. This root problem is human population growth. In the United States and Canada, it is also the issue of per capita resource consumption. If we do not address and reverse these trends, it will be difficult, if not impossible, to ensure that our generation's grandchildren will ever see anything remotely resembling healthy natural ecosystems anywhere in the Great Lakes–St Lawrence watershed: rivers teeming with aquatic life, riparian forests with wild predators and grazers, and diverse bird populations.

Statistics released by the U.S. Census Bureau show population growth in the United States has climbed from a negative rate throughout much of the 1970s and early 1980s to a major new baby "boomlet" in the 1990s. The increased growth rate is a combination of rising birth rates and greatly increased immigration, both legal and illegal. In Canada as well, increased rates of net in-migration contribute to population growth. World population, meanwhile, continues to expand exponentially.



Given the rates of resource depletion and consumption by residents of the United States and Canada (who consume natural resources at rates conservatively estimated at more than ten times the rest of the world), even relatively low population growth in North America has a disproportionate impact on the environment, locally and globally.

Lester R. Brown, president of the Worldwatch Institute, provides a graphic image of this increase: "Our generation is the first to witness the doubling of world population during a lifetime. Indeed, everyone born before 1950 has seen world population double."[5] The United Nations Population Division estimates that human population reached 6 billion in 1999, and will reach 7 billion by 2010 and 10 billion by about 2050.

The problem of population growth is greatly responsive to political leadership. It was no fluke that the decline of the birth rate in the United States through the late 1960s and early 1970s coincided with major political attention to the issue from administrations as divergent as those of Nixon and Carter. (See, for example, the *Report of the Commission on Population Growth and the American Future*, 1972, or *Global 2000*, compiled by the Carter Administration and published in 1980.)

Current growth rates are greatly influenced by the outright hostility to the issue by the Reagan administration and the dearth of political leadership of the Bush administration. As Brown notes, the 1980s "turned out to be the lost decade, one in which the United States not only abdicated its leadership role, it also withdrew all financial support from the United Nations Population Fund and the International Planned Parenthood Federation."

While government can provide non-coercive policy leadership with speeches, tax incentives, subsidization of birth-control research, contraceptive information distribution, and other methods, ultimately in a democratic society the choice to have a child is rightfully left to individuals.



## Building a Watershed Movement

The framework for grassroots watershed conservation exists in the three thousand or more river guardian groups across the continent. The challenge is to work with international, regional, provincial, state, and citizen groups to foster a cohesive movement for river and watershed protection. This means recruiting and empowering leaders. It means building the local citizen organizations so that they are capable of carrying out campaigns. It means linking up all these leaders and organizations so that they can work together for the common goal, to stem the tide of river deterioration and forge new tools for watershed conservation. It also means building the personal relationships where we live, with our neighbours and businesses, river conservation colleagues, and key decision-makers at all levels of society.

In two respects this is easy. First, we are a continent dependent on rivers, reservoirs, and lakes for our drinking water. More than 85 percent of all U.S. citizens get at least some of their water from rivers, and more than 75 percent of Canadians rely on rivers and lakes for their drinking water as well. And even those reliant on groundwater for drinking ultimately share the same concerns.

Secondly, watershed approaches also provide unique opportunities for improved environmental justice throughout North America. Rivers are so intricately woven into the fabric of urban and rural society that they affect both the wealthy and poor.

Poor people, though, are most at risk when a river becomes degraded. They rely on rivers for drinking water to a larger extent, rely on catching contaminated fish for a large part of their food supply, and therefore are more exposed to pollution and contamination. Watershed action, with an emphasis on empowering urban grassroots organizations and poorer rural communities, can become a major tool for improving human health and increasing equity in North American environmental policy.

Long-term success for watershed protection and restoration will hinge on our ability to mobilize existing river guardian organizations, reaching out to new constituencies in the inner cities, businesses, governments, and environmental organizations in expanded watershed protection efforts in the future. Some of these efforts are underway. For example, River Network, a national nonprofit organization founded in the United States in 1988, provides a wide variety of organizing efforts, technical assistance, and policy help to local citizens throughout the country from their headquarters in Portland, Oregon.



Federal agencies including the EPA, U.S. Forest Service, Fish and Wildlife Service, and Bureau of Land Management have all adopted new programs for ecosystem and watershed restoration and management, as have the Canadian federal departments of Environment, and Fisheries and Oceans. The immediate challenge for the river conservation movement is to coordinate, connect, and expand the grassroots constituency as fast as river science and public policy have developed in the last few years.

## "The Brook"

How do we most effectively use the data gathered on the Canadian rivers, or that presented by the Nature Conservancy and the American Fisheries Society? Site-specific threats literally occur by the thousands. To answer this question, let's look at the recent history of one relatively small brook.

Thirty years ago when a child regularly hiked and fished and swam at "the Brook" most of its several-mile length was fairly well insulated from direct human impact. Turtles, salamanders, water striders, and brook trout were easily found. Then there were only about three houses within three hundred yards of the brook; but twenty-five years later there were over forty in just the upper reaches. By this time, in the summer, little or no water was left in this stretch. It was silted and dried up. No fish remained, no turtles, no salamanders – nothing much for little kids to play with or to discover. An old gravel pit had expanded and diverted what was left of the stream. There were no longer bitterns in the cranberry bog near "Bittern Lane," and the pond nearby drained in a trickle to the brook's confluence with a nearby river. A sad story of a lost resource.



There may yet be a happy ending. After a brief visit to that town in 1993, Russell, a seventy-one-year-old family friend and avid fisherman, told some new stories about "the Brook." Russell spoke of some low-key natural restoration that has taken place in the lower stretch below the new subdivisions. Beaver have returned above the gravel pit, damned the brook, and created new fishing holes. Russell fished to his daily limit in the beaver pond twice in the summer of 1992. The return of "the Brook" has proven to be a true story.

This anecdote reveals the tremendous resilience of rivers, streams, and watersheds – if they are given a chance. It also illustrates the possibility of restoring the brook's entire length if we change the way we use resources and interact with natural systems. In this case, the decline of trapping allowed the beaver to return to the drainage basin, thereby restoring some of the brook's natural functions. Restoration of other reaches will require human intervention, but the opportunity is there.

"The Brook" and the thousands of others like it represent a challenge in the new century – a challenge to increase river protection to a new level of political awareness, protection, and effective restoration. We need to change the way we treat our rivers, and address the landscape and our watersheds in new and innovative ways. We must do this, regardless of whether our ecological address is "the Brook," or the Black, Grand, or Cuyahoga rivers, or the Grand or Ottawa rivers, the St Francis, the St Maurice, the Saguenay, Rivière du Loup, or the Gatineau, or the Bad, Wolf, Fox, Menominee, Muskegon, Manistee, Maumee, or Genesee rivers, or the Raquette.

Restoration of the innumerable brooks and rivers of this great continent is our collective challenge. This is where the myriad efforts of citizen activists, volunteer monitors, restoration ecologists, visionary politicians, progressive businesses, and interested citizens will go now and in the future. To ensure the health of streams, rivers, and lakes, we must also focus on the protection of headwater, upland, and wetland areas. Ecosystem integrity and the well-being of our precious waters demand that we address environmental issues from a holistic, watershed perspective.



Facing page: Lady Evelyn Smoothwater Provincial Park. Photo: Frozenrope

Section 2

# FINDING A VOICE

*"If you want to understand the moods of a watershed, you have to walk its shorelines.*
*You must feel the sand, pebbles, and stones of the beach under your feet.*
*Then you need to walk the riverbanks, pushing through the tangled undergrowth and climbing over the deadfalls that hide the rabbits and grouse.*
*Sometimes you have to get your feet wet."*

– Michael Keating

Photo: Michael Keating

*"I have always lived in a town or city that was built on a great lake or river. Over time I learned that the waters have many moods. Like humans, they seem to have both a bright and a dark side. On good days, the light glints off the waves, and you can swim in the warm bays and let the currents carry you down the rivers. On bad days, the wind rises, and big lake boats are tossed around like toys in a bathtub . . . Whenever I go back to my home town, I am drawn to walk the shoreline, even on the foulest days. This is how I keep in tune with nature and up to date with news of the ecosystem."*

*–Michael Keating*

Photo: Frozenrope

# Revelations

## *The Evolution of an Environmental Ethic and Career*

*Michael Keating*

*"When I become a reporter in 1965, the word environment hardly appeared in the news. It had only been three years since Rachel Carson's book,* Silent Spring, *had begun to awaken people to the threat from toxic chemicals and helped trigger the modern environmental movement."*

I GREW UP IN A SMALL LAKESHORE TOWN. My first memory of water is one of a great golden expanse of sand meeting the sharp blue dividing line of the lake. I can still feel the July sun burning my shoulders and the cool water on my ankles where the blood flows close to the skin.

I have always lived in a town or city that was built on a great lake or river. Over time I learned that the waters have many moods. Like humans, they seem to have both a bright and a dark side. On good days, the light glints off the waves, and you can swim in the warm bays and let the currents carry you down the rivers. On bad days, the wind rises, and big lake boats are tossed around like toys in a bathtub. When autumn storms coincide with high water levels, the waves can roll across beaches to batter homes off their foundations. Every fall when the winds blew hard from the west, I could hear the constant roar of the lake pounding on the shore not far from our house. I remember standing on that beach, watching an endless chain of huge waves that seemed to be rolling out of a bank of purple clouds on the horizon. The water was cold and steel grey. Swimming season was over.

If you want to understand the moods of a watershed, you

have to walk its shorelines. You must feel the sand, pebbles, and stones of the beach under your feet. Then you need to walk the riverbanks, pushing through the tangled undergrowth and climbing over the deadfalls that hide the rabbits and grouse. Sometimes you have to get your feet wet. Whenever I go back to my home town, I am drawn to walk the shoreline, even on the foulest days. This is how I keep in tune with nature and up to date with news from the ecosystem. While walking my favourite beach, I found zebra mussel shells, tucked one into the pocket of my parka, and made a mental note that the latest invading species had reached this part of the watershed. Later on the docks at the harbour I saw the same small, striped shells clinging to the bottom of the fish tugs hauled up for the winter to keep them out of the ice floes.

The town was built where a large river empties into a lake. This area, rich in fish and game, had been settled for thousands of years before the first Europeans arrived at the river mouth and set up small fishing and lumber camps. Here as elsewhere across the land, settlers cut and burned many of the forests to create farms. To those immigrants, the great stands of pine, maple, oak, and beech that reached to the shorelines seemed vast and impenetrable.

Just off the town lies a large, low, stony island. For a century and a half, boats have taken shelter from storms behind it. Several of them foundered on its reefs and never left. From the shoreline two docks reached like outstretched arms toward the island. Built in the late 1800s, they formed a harbour for ships from around the lakes. The docks became ribs in a great railway network that reached out from the burgeoning cities of Montreal and Toronto to spread across the basin in the second half of that century. The town became one of the railheads, and it prospered from shipments of timber and fish. Over a few decades the vast forests were cut and lumbering came to an end, replaced by farming and furniture factories. Then, it became more economical to ship products such as furniture from the new factories by rail and later by truck. The docks fell into disrepair, like trees decaying on the forest floor. When swimming, we would find the great square spikes that held the timbers together, half buried like pirates' treasure on the sandy lake bottom.

I spent many an afternoon walking the ruins of the old docks, one made of rock and timbers, the other of concrete filled with boulders. From the concrete dock I watched the last lighthouse-keeper rowing his skiff back to the island on a choppy lake that reflected the setting sun. I did not know this was an historic sight, since he would not be replaced. The lighthouse tower remains standing, its beacon run by machines, but the light-keeper's house of white painted stone has fallen into disrepair. We would stand on the beach with war surplus binoculars that had range-finding lines across one lens and count the holes appearing in the roof. On windy days we could see ducks, geese, and gulls bobbing in the lee of the island, which, abandoned by people, had become a bird sanctuary.

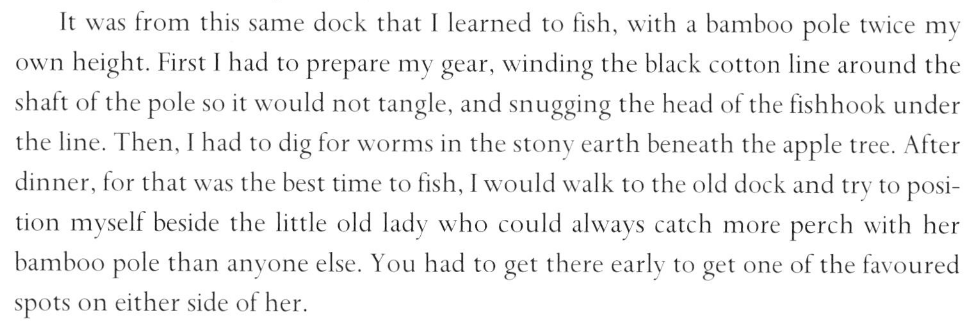

It was from this same dock that I learned to fish, with a bamboo pole twice my own height. First I had to prepare my gear, winding the black cotton line around the shaft of the pole so it would not tangle, and snugging the head of the fishhook under the line. Then, I had to dig for worms in the stony earth beneath the apple tree. After dinner, for that was the best time to fish, I would walk to the old dock and try to position myself beside the little old lady who could always catch more perch with her bamboo pole than anyone else. You had to get there early to get one of the favoured spots on either side of her.

Some days I would just lie on the concrete walls and watch the water below. On a good day you could see a school of bass gliding above the sandy bottom. Once a huge fish, likely a sturgeon, was thrashing in the shallow waters between the docks. Probably no one alive today remembers when the trout ran so thick in the river mouths that you could club them with an axe and carry home your dinner. Few if any living people remember the catches of sturgeon half again as long as a full-grown man and twice as heavy. Sturgeon were once monarchs of the lakes. But nineteenth-century fishermen considered them a nuisance because their armoured bodies tore nets, and in some cases they burned these fish like cordwood on the beaches. Sometimes called living fossils, because they resemble fish that lived in the time of the dinosaurs, sturgeon are now considered a vulnerable species. The increasingly rare sight of one of these giants racing through shallow water like a submarine is enough to make the pulse jump.

Now the dock I used to walk as a child, the dock I learned to fish from, is nothing but a pile of boulders beneath the waves. Even the railway is gone, leaving only the occasional steel spike rusting in the weeds beside the railbed as a childhood souvenir. No freight-carrying schooners shelter behind the island. They have been replaced by yachts, windsurfers, jet skis, and outboard motorboats towing water-skiers.



As a child I began to realize there were more problems in the fisheries than my sorties with hook, line, and sinker. The discovery began as a typical family outing. We piled into the old Plymouth and headed to the ruins of a bridge washed out by a spring flood a few years earlier. Standing at the river's edge, we watched people tending the lamprey traps. Local residents were repelled by the idea of eating something that looked like a snake, so the canned meat was shipped to the big cities. While we crouched where the water ran between the stones, a man bent down beside me, a lamprey writhing in his hand. The slippery creature was as long as his arm. To illustrate the power of its rasping, suction-cup mouth, he held this eel-like fish just behind the blunt head and plunged it into the shallow water and against the river bottom. Then he triumphantly lifted the lamprey out of the water, with a very large rock firmly attached to its mouth.

I can never forget the sense of power in that lithe beast. Anything that could lift a rock that weighed far more than itself could obviously attack any fish. Over-fishing was a problem, but those powerful mouths with razor-like tongues helped to suck the life out of the once-great fisheries through Lakes Huron, Michigan, Superior, and Georgian Bay. Governments decided to stop the lamprey, both by building dams to prevent the creatures from going upstream to spawn and by poisoning the larvae of those that slipped by the defences. Where I once watched the lamprey traps, a large concrete dam now spans the river. The Pacific salmon that were stocked in place of native lake trout can migrate up the river through fish ladders – a series of platforms – but the lamprey are blocked by this watery puzzle.

Like all my friends, I grew up spending my summers sprawled on the beach between plunges into the icy lake. Once the early summer sun began to warm the deep, cold waters, we were swimming almost every day. Every so often someone who swam in the foamy water near the old docks would get what we called fisherman's itch, a reddish skin rash that cleared up in a day or so.

When the lake was too rough or cold, we would sometimes swim in the warmer, more sheltered waters of the river just north of the beach. We were told to keep our heads up and never to swallow the river water. Even so, some of us, especially those who dove into the river from the docks, came down with a stomach, eye, or ear infection. This was an introduction to the realities of water pollution. The river was full of bacteria from the untreated discharges of sewer pipes, septic tanks, and cattle manure upstream. Even then, in the early 1950s, sewage treatment was unknown in most towns and cities. People discharged their wastes into septic tanks, many of which seeped into creeks and rivers. Sometimes the household drains emptied into storm sewers that flowed directly into the rivers and the lakes.

When I was young, the town's drinking water intake was close to the river mouth. Each spring when the ice went out, erosion from upstream farms would fill the river with silt, overwhelming the ability of the water treatment system to remove the dirt. Most years the tap water was so muddy we didn't drink it. We would go down to the basement and drag up our one-gallon glass jugs. If we were lucky, the neighbours' well was full, and we could get water from them. Sometimes their well was dry, so we would load the bottles into the car and drive down to a spring near the river. A galvanized iron pipe came out of the ground, took a right-angle bend, and discharged a small but steady stream of clear water. It was safe to drink but foul-tasting from sulphur. The choice between silt and sulphur was a difficult one, but the tap water would clear after a week or so, and we could avoid the problem for another year. Since then the town has installed a modern filtration plant, but the local stores have taken to carrying bottled water, which sells briskly.

I grew up close to what I thought was a big river, an important tributary of the

Great Lakes. Not until I saw the Niagara River, and finally the St Lawrence, did I realize a river could be so big as to be majestic. It was 25°C below zero, and four of us were packed into an old Chev, driving to the Carnaval, Quebec City's famous mid-winter festival. As we followed Highway 2 past Kingston, a wide, frozen expanse of silver appeared on the horizon. While we drove, the St Lawrence played hide and seek behind the trees and the white frame houses of United Empire Loyalist country.

It was a bitterly cold and sunny day when we stood on the ramparts of Quebec and looked out over a broad stretch of ice-choked river. Far below we could see teams of men, hardly more than dots on the ice, propelling wooden rowboats across the ice floes from Quebec to Lévis on the far side of the river. They leaned on the gunwales and pushed off the treacherous ice with their rubber boots. As we watched the famous ice boat races, we noticed that around us men kept tipping bright-coloured plastic canes into the air. They were filled with liquor, or "antifreeze," as one rosy-cheeked gentleman informed us.



Later I would discover a more tender side of Le Fleuve. On a sunny June afternoon I was driving across the great arc of a bridge that links the north shore to the Île d' Orléans. This large island with its woods, farms, and villages neatly splits the river just downstream from Quebec City. With friends from the island we explored the boulder shoreline and watched ocean-going freighters sliding by. As our gaze followed them upstream, we saw the towers of Quebec City silhouetted against the setting sun. It is at Quebec, the last big city to draw its drinking water from the Great Lakes–St Lawrence ecosystem, that you get a real sense of the history, power, and grandeur of this mighty river.

## Exploring by Canoe and by Ski

Discovering the canoe was like finding a magic carpet with which to explore the watershed. You are on the water and of the water. You can skim across lakes pushed by the wind, bounce down rapids, poke slowly under the willow trees, and push through the lily pads and reed beds. The voyages began with a friend whose father owned a boat livery. When business was quiet, we would slip one of the long green cedar and canvas Chestnut canoes off the wooden dock and into the warm waters of the river. We spent summer afternoons on this winding, sandy stretches, learning to steer with J-strokes as we paddled past swifts and fishermen in motorboats anchored to the river bottom. With a canoe we could silently drift up to old sun-bleached logs, watching to see how close we could get before the painted turtles would raise their heads, then slide into the water.

As I got older it was one canoe or another that let me explore more and more of our watershed. The patterns of land and water change subtly as you leave the clay and sand of the lower lakes and St Lawrence valley and head north into Shield country. The rich, deciduous forests, with their rounded skylines of maple, birch, and poplar, give way to a more ragged horizon of red, white, and jack pine. This is succeeded by the tall, slender black spruce spires of the boreal forest. Here you are on the Precambrian Shield with some of the oldest rock on earth. You look for patches of moss or deposits of pine needles on which to plant a tent. In the evening when the wind dies down, the loons cruise the waters off your campsite, busy with their own fishing expeditions.

The rivers that spread out like arteries and veins carry the history of our landscape. On the French River I learned to paddle in rapids such as the historic Blue Chute before heading off to try other whitewater rivers. Where we played in our tough, synthetic canoes, aboriginal people and European fur traders over the centuries weighed the risks of running the rapids. Some failed to make it, and across the country the loads from those canoes lie at the foot of many a rapid. East of the French, in Quebec, lies the Bazin, a tributary of the Gatineau, which in turn feeds the Ottawa. Rivers were the highways of commerce until a century ago, and tall timber, especially pine, was the coin of the realm. Logs were floated to market down the rivers in great spring drives. When we paddled the Bazin, the log drives were nearly over, and most wood was trucked to pulp mills and sawmills. This was safer for the loggers, many of whom were injured or killed trying to clear log jams, and in some respects was easier on the environment. The log flotillas ripped out spawning beds and left long stretches of river bottoms covered with bark and wood that smothered life. But there were still traces of the old days. At times we were paddling in a sea of pulp logs, pushing our way through. When we arrived at one big set of rapids, the river was blocked as far as the eye could see by a massive log jam that had never been cleared.

To the west of the Bazin-Gatineau watershed, the Dumoine River flows out of western Quebec into the Ottawa, cascading over one ledge after another. This pool-and-drop type of river creates a paradise for modern whitewater canoeing, but the rapids used to be seen as just one more obstacle to getting the great pine logs out of the forest and down river. About two days' paddle downstream from the first set of rapids there is a long series of cascades known as Grand Chute. If you look closely into the bush while walking the west bank, you can see the remains of timber and stone cribs that once supported a flume to carry logs past this cataract.

Canoeing has become a popular form of recreation in recent decades. On the Au Sable River in northern Michigan it was not a log jam but a canoe jam that prevented us from launching for several minutes. So many people had decided to get away from the crowds of southern Michigan on that spring afternoon that we had to line up to get onto the water. Yet, a couple of hours later, after the rain showers had begun and we had moved downstream, the river was quiet. Our canoes were trying to glide past fly fishermen in hip waders without disrupting either them or their quarry, the trout.

In Algonquin Park I went looking for wild rivers but also discovered wetlands. Here as elsewhere, many rivers start in swamps, the natural sponges that keep releasing cool water even during the long, hot days of summer. At best we humans, who no longer feel comfortable walking on soft, wet soil, dismiss wetlands as places where there is too much water for a picnic and too little for swimming. In legends and movies we are taught to think of the bogs, fens, swamps, and marshes of this world as dark haunts of demons and other evil creatures. We often consider them as wastelands which should be drained to make room for farms and cities.

However, this meeting ground between land and water is where most forms of life want to be, and wetlands are among the richest of all ecosystems. Their warm, shallow waters are up to eight times more efficient than a wheat field at turning the sun's energy into plant life. Marshes, their myriad channels choked with plant life, slow the flowing water, trapping sediment and contaminants such as fertilizers and pesticides. The same "drag effect" of vegetation on flowing water makes wetlands natural flood barriers, capturing runoff during the spring snow melt and after rainstorms. But when you are up to your knees in mud, dragging a heavily loaded canoe by a rope, and swatting mosquitoes, it is hard to remember all the benefits.



Once the river moves out of its headwaters and carves a clear route, you are in a world of two directions – upstream and downstream. You want to be moving downstream with the current, sometimes paddling rapids, but often portaging past the rocky stretches. Then the canoe becomes a giant carapace over your head, trying to push your shoulders into the ground. Finally the river has run its course and empties into a lake. On the open water the world becomes circular once again. It is on the lakes that you feel most vulnerable in a canoe, and you most keenly sense the power of nature.

Paddle down the French River to the islands of Georgian Bay, and you gradually depart the closed world of the river to the ever more open one of the bay. From the seat of a canoe you experience the vastness of the sky and the lake and feel very fragile in your small boat. The sense of being a tiny piece of flotsam is greatest when paddling the shores of Lake Superior. Before the trip, we read the warnings of great storms that can sweep across its vast reaches and thought of how quickly one of them sank the *Edmund Fitzgerald*, an ore carrier as big as an ocean freighter. Each day we watched the sky for signs of a storm, or a fog bank that could isolate us from shore. We constantly scanned the water ahead for splash marks that showed hidden rocks just under the waves. Each afternoon we left the open waters and paddled into a deep, sheltered bay to camp on a sand beach and listen to the warblers as they fluttered across the surface at dusk.

Superior is a giant inland sea, as frigid as the North Atlantic. Only the shallow

waters at the head of a bay can be warm enough to allow a swim before cooking dinner over a driftwood fire. This region is as close as one can get to an untouched ecosystem in the whole Great Lakes–St Lawrence drainage basin. Cold climate and poor soils encouraged the millions of travellers who passed this way to keep moving. Over 90 percent of Superior's watershed is still forested. This is the cleanest of the lakes, and when paddling far away from the few towns and cities along its shoreline, we drank the water directly from the lake. It is clean enough for drinking but no longer pristine. Toxic chemicals from pulp and paper mills around its shores have entered the food chain. So have chemicals that are transported on the winds and have accumulated in the fish.



Skiing became my second magic carpet across the length and breadth of the watershed. I learned to ski on the same beach where I spent my summers, with sand dunes becoming tiny downhill runs. On sunny days we would seek out the wind-crusted slopes and learned how to crouch and glide on our solid wood skis with their beartrap bindings. One year the lake froze so hard that the island became united with the mainland, and for a few weeks the ice was thick enough that we could ski to the foot of the lighthouse. In one spot the lake froze smooth and glassy, and some teenagers pushed the snow aside to create a temporary hockey rink.

Often cold northwest winds off the lake drove us from the beach and into the woods. Here we were able to glide over ponds and deadfalls, exploring marshy areas that are virtually impenetrable the rest of the year. In the dead of winter only a few fast-flowing creeks remain open below the beaver dams. Over time, I began to range farther afield with both downhill and cross-country skis. While canoeing gives you a sense of how water moves through the valleys, skiing shows you the lay of the land, revealing how watersheds are defined by heights of land. In Ontario's Beaver Valley or Devil's Glen you ski down toward the valley floor where the snow will someday melt into the waiting rivers. Where the Niagara Escarpment faces Georgian Bay, such as west of Collingwood, you look out over the frozen waters of the Great Lakes system.

In northern Michigan and outside Thunder Bay, ski hills rise up to dominate the landscapes around them. Once, in the Agawa region above Sault Ste Marie, I got off the Algoma Central train in the middle of the bush and skied along a creek until I saw smoke rising from a lodge. That wilderness lodge, a mecca for trout fishers in the spring, was nestled among the pines on an island in a small inland lake. From that base you could climb the rugged hills that inspired painters in the Group of Seven. On one side was the frozen expanse of eastern Lake Superior, on the other a chain of inland lakes and rivers among the forested hills.



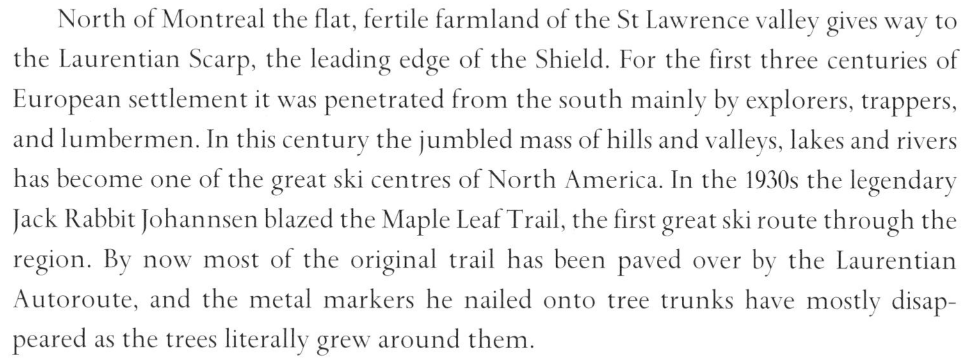

North of Montreal the flat, fertile farmland of the St Lawrence valley gives way to the Laurentian Scarp, the leading edge of the Shield. For the first three centuries of European settlement it was penetrated from the south mainly by explorers, trappers, and lumbermen. In this century the jumbled mass of hills and valleys, lakes and rivers has become one of the great ski centres of North America. In the 1930s the legendary Jack Rabbit Johannsen blazed the Maple Leaf Trail, the first great ski route through the region. By now most of the original trail has been paved over by the Laurentian Autoroute, and the metal markers he nailed onto tree trunks have mostly disappeared as the trees literally grew around them.

Heading east from Montreal, you can follow the Autoroute des Cantons-de-l'Est, then swing south on local highways. Here, in the Eastern Townships, the hills rise suddenly from a flat terrain, foreshadowing the Appalachian Mountains that reach far south into the United States. Across the border in Vermont the ski peaks loom higher and higher – Jay Peak, for example, and Mount Mansfield with its famous Stowe resort.

For a few months of the year nature and water seem to stand still. But when you look down from any ski hill in spring, you can see the hydrological cycle starting up for another season. At the bottom of each gully, a stream is starting to run. While the snow lingers on the north-facing slopes, the farm fields below are turning brown. As the water runs down the hill and across those fields, it is picking up everything from soil particles to bacteria to agricultural chemicals, and carrying them into the rivers and lakes. This diffuse, non-point source pollution is an important source of contamination, but because it is so scattered, it is difficult to control.

On the shoreline, spring is the time of the singing ice. As the warm spring sun melts the ice banks, they become honeycombed, and the disintegrating mass of ice crystals gives off a faint ringing sound. It is the in-between time of the year, when there is too little snow to ski and too much ice for boating. It is a time to wait for the ice floes to thin enough so the smelt can once again run in the streams, and anglers can wait for them in the night with flashlights ready and dip-nets poised.



### Becoming an Environmental Journalist

When I became a reporter in 1965, the word environment hardly appeared in the news. It had only been three years since Rachel Carson's *Silent Spring* had begun to awaken people to the threat from toxic chemicals, helping to trigger the modern environmental movement. Journalism traditionally dealt with the natural world through the lens of adventure travel or outdoors writing, which focused on activities such as hiking, hunting, fishing, and bird-watching.

I stumbled on my first pollution story in 1966 when a friend called. A year earlier, after we graduated from journalism school, I went to work for a newspaper and he for a government water pollution control agency. One day he phoned, saying, "I want you to come to Niagara Falls to meet a congressman who has something interesting to say about pollution." We met on an observation deck periodically blanketed with mist rising from the falls. The congressman was handing out tiny golden lapel pins shaped like fish, and saying that if the United States and Canada did not do a better job of controlling water pollution, the fish would be dying. In particular, he talked about a threat to Lake Erie just upstream from the falls.

It was hard to grasp the idea that an entire Great Lake could be so polluted that parts of it had become lethal to fish. The complex process was linked to phosphorus, one of the more common and seemingly innocuous elements around us. Phosphorus, an essential nutrient for plants and animals, is used to make fertilizers, soften water, light matches and, as phosphoric acid, give the bite to cola drinks. When detergents became popular some decades ago, it was phosphorus in the form of phosphate that helped get clothes whiter than white. This introduction of phosphate detergents coincided with the baby boom and the wave of immigration that followed the Second World War; the resulting rapid rise in population in the Great Lakes region led to more detergent use. At the same time farmers were using fertilizers to increase production, and house-proud citizens of the economic boom spread ever-increasing amounts of fertilizers on lawns and gardens. By the 1960s there was a great accumulation of phosphorus from a combination of human sewage and wash water plus rainwater carrying excess fertilizer off farm fields, lawns, and gardens. It amounted to thousands of tons of the stuff flowing into the heavily settled Lake Erie and Ontario basins each year.

Phosphorous, as any gardener knows, is an important plant food appearing as the second of three numbers printed on a bag of fertilizer. In Lake Erie the flood of phosphorus-rich sewage and agricultural runoff fertilized the growth of algal blooms so massive that they painted the water bright green. As algae died and sank to the lake bottom, the bacteria that consumed them used up the tiny oxygen bubbles in the water, creating so-called dead zones on the lake bottom where oxygen-breathing life would suffocate. In the media the story was simply that Lake Erie was "dying," though many scientists pointed out that most of the lake was actually *too* "alive" with plant life. For the average person the result was a lake full of rotting weeds that gave a foul taste and odour to drinking water and piled up knee-deep on shorelines. On one hot summer day in the late 1960s, I put the top down and drove to a Lake Erie beach in anticipation of a refreshing swim. The water was the colour of pea soup, and my feet soon disappeared in this murky green liquid. I remember the feel of strands of plant life brushing against my ankles as I felt for the bottom. I held my nose, waded part way in, then retreated to the showers with chlorinated but clear water.

Phosphorus was far from the only troubling pollutant. The Great Lakes and

St Lawrence region, with tens of thousands of shops and factories, including petroleum refineries, chemical plants, steel works and pulp and paper mills, is one of the world's great industrial centres. The Sault Ste Marie locks handle more tonnage per year than the Suez and Panama canals combined. Canada's chemical valley lies along the St Clair River, and one-quarter of U.S. chemical companies are in the Great Lakes region, with much of that on the Niagara River. One-fifth of U.S. industry and half of Canada's manufactured goods come from the Great Lakes basin. From Quebec City to Duluth, this region is home to around 45 million people. It includes Toronto and Montreal, along with such major American centres as Buffalo, Cleveland, Detroit, and Chicago. Over the years the towns, cities, and industries have dumped millions of tons of wastes into the water. In the 1960s the Rouge River at Detroit was so dirty that twenty thousand ducks were killed on a single winter's day after they landed on an oil slick. This kind of pollution was common in a society that ran on petroleum and was careless with its wastes. In 1931 a slick caught fire on Toronto's Don River, damaging a bridge. But the incident that really caught the public's attention was the 1969 fire on the oily surface of the Cuyahoga River at Cleveland. In this case, fireboats were called in to help quell the blaze.

The United States and Canada have been wrestling with the problem of Great Lakes pollution for more than a century. In the 1800s two out of every thousand people died each year from typhoid, cholera, or similar illnesses contracted from drinking water polluted with human or animal wastes. In 1909 the two nations signed the Boundary Waters Treaty. Though mainly intended to prevent disputes over border waters, the treaty contained an early reference to the need to deal with pollution: "boundary waters and water flowing across the boundary shall not be polluted on either side to the injury of health or property on the other." To make the agreement work, the two countries created the International Joint Commission to investigate and make recommendations on boundary water problems. It is a high-level group: three members are appointed by the U.S. president and three by the Canadian prime minister.

In 1912 the IJC was asked to investigate pollution in the Detroit and Niagara rivers. Six years later, after the hiatus of the First World War, the commission's final report stated that the situation was "generally chaotic, everywhere perilous, and in some cases, disgraceful." The major problem was bacteria, because sewage systems, where they existed, were simply pipes that carried wastes to the nearest watercourse. The solution to pollution, people thought, was dilution. It worked up to a point, but as the populations grew, the capacity of the lakes and rivers to purify the wastes was surpassed. People were saved from themselves by the invention of drinking-water chlorination. This highly toxic gas is still bubbled through the drinking water in most municipalities to kill bacteria. A small amount of chlorine, called a residual, is left in the water to kill any bugs that get into the pipes along the way.

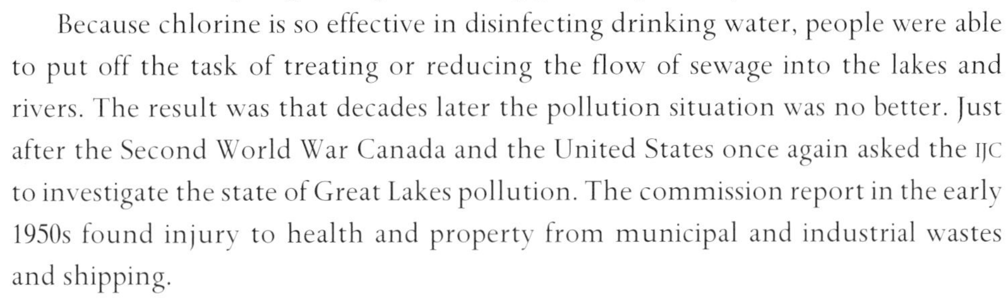

Because chlorine is so effective in disinfecting drinking water, people were able to put off the task of treating or reducing the flow of sewage into the lakes and rivers. The result was that decades later the pollution situation was no better. Just after the Second World War Canada and the United States once again asked the IJC to investigate the state of Great Lakes pollution. The commission report in the early 1950s found injury to health and property from municipal and industrial wastes and shipping.

In addition to bacteria, the commission found a growing load of synthetic chemicals, the legacy of the chemical boom that followed the war. Pesticides such as DDT were considered miraculous inventions because they killed unpleasant insects like mosquitoes and house flies on contact. (As a child, I used to chase flies around the house with a spray pump loaded with the stuff.) But chemicals like DDT do not break down in the environment, and build up in the food chain to levels that kill wildlife. By 1959 they were detected in Lake Erie and four years later were recorded in the St Lawrence River ecosystem. Since Rachel Carson's warnings in *Silent Spring,* there had been growing concern about the effects of chemicals on wildlife such as birds, but humans seemed relatively unaffected.

That concern was strengthened late in the winter of 1970, when scientists revealed that many fish from Lake St Clair and some from the western end of Lake Erie had high levels of mercury in their flesh. The mercury, which could cause nerve damage, had come mainly from Canada's chemical valley upstream on the St Clair River. For me, and for millions of others who lived downstream from the chemical discharges, it was a rude awakening to the reality of toxic chemicals in the food chain. The previous summer I had been fishing for bass on Lake St Clair and eating the barbecued fish. Was I contaminated? Later in the spring of 1970, I found myself in a government laboratory, photographing scientists analysing fish for mercury. Over time I learned that mercury was just one of more than 360 contaminants in the ecosystem around me, including the water, sediments, fish, animals, waterfowl, and humans.



A scientist would later tell me that he, I, and everyone else in the industrial world carry trace amounts of a wide range of chemicals in our bodies. The list includes fire retardants, electrical insulating fluid, insecticides, weed killers, and a number of chemicals used by industries to do such things as clean the grease off metals and make other chemicals. Natural pollutants such as phosphorus are eventually broken down into nutrients or harmless substances by the action of sunlight and bacteria. Synthetic chemicals such as DDT, PCBs and dioxins are created by combining petroleum and chlorine under heat. One of the characteristics of organochlorine chemicals is that they do not break down easily in the environment. Another characteristic of these persistent pollutants is that they are highly soluble in fat, so they collect in living cells in the process of bioconcentration.

Drawing diagrams on a napkin, the scientist explained how chemicals typically enter the base of the food chain at the level of the plankton. Sketching out ball-point images on the soft paper, he showed how the plankton are eaten by small fish, that are in turn eaten by ever-larger fish, and the package of chemicals in each body is added together. That is biomagnification. PCB levels in the eggs of bald eagles nesting along the shores of Lake Erie have been measured at up to 25 million times higher than those in the surrounding lake water. Humans, I was told, are a top predator, near the upper end of the food chain.

More than a decade after meeting the congressman on the banks of Niagara Falls, I was back for an even bigger story. Long the honeymoon capital, Niagara had also become one of the chemical centres of the world. Since the turn of the century, chemical industries had been drawn to this region, particularly on the American side, by the endless flow of water and cheap electricity from the falls. As the industry helped to create the chemical era that followed the Second World War, it also produced huge amounts of wastes. These discarded chemicals, some with names that only an engineer can decipher, were generally dumped into the river, or buried in pits which began to seep into the waterway.

In 1977 the problem of chemical wastes literally bubbled to the surface. It happened in a quiet, suburban neighbourhood in the southern portion of Niagara Falls, New York. Years before, an abandoned shipping canal, Love Canal, had been filled with chemical wastes. It was sealed up, the land sold, and homes built over the pit. All was peaceful until a series of record rains caused the canal to overflow its underground banks, floating chemicals up and onto the surface. People were driven from their homes by the stench of toxic chemicals leaking into their basements, while children and animals were chemically burned when they played in contaminated puddles. The neighbourhood nearest the canal was fenced off and declared a federal disaster area – the first time for a disaster created by human hands. Within two years about 850 families were evacuated and their homes boarded up.

The human impact of that pollution was driven home one day in 1982 when another journalist in Niagara Falls, New York, called me. "You should come here today," he said. "They're going to start bulldozing the houses." A few hours later I arrived with a photographer in time to see two giant yellow bulldozers belching diesel smoke as they rumbled toward frame buildings which had been boarded up since the evacuation. I remember Dolores Frain crying bitterly as one of the bulldozers smashed her home of twenty-three years off its foundations and crushed it into the chemical-laden dust, "It's just as if you are burying one of your own." The other journalist suddenly turned to me, wrinkling his nose. "Do you smell those chemicals?" he asked.

Workers wearing filter masks were hosing water onto the site to settle the dust, but within minutes the stink of chemicals was everywhere. We held our breath, jumped into the car, and drove out of that hell-hole of broken dreams and toxic waste.

For many years Love Canal was the most infamous toxic waste site in the world, and a symbol of the industrial dream gone awry. It was not the only dump on the Niagara frontier. After governments probed more deeply, they found more than 160 toxic waste dumps close to the river. With news that tons of chemicals were leaking or being discharged into the river every day, many of the millions of people downstream became concerned about their drinking water.



When the first European explorers sailed up the St Lawrence and paddled the Great Lakes in birch-bark canoes, one could dip a cup in nearly any lake or river and drink the water with safety. If you tried that now on any city waterfront, it would be an open invitation to a whole shopping list of microbial infections, including those that ravaged previous generations. The fear of getting ill or even dying from contaminated drinking water virtually disappeared with the advent of chlorination early in the twentieth century. Fears of what might lurk in a glass of water came back in a new form after the 1980 announcement that TCDD dioxin, the most toxic synthetic chemical known, was seeping into Lake Ontario from Niagara dumpsites. What had been a nascent fear about the safety of drinking water blossomed into near panic, and a generation that had trusted tap water almost implicitly suddenly began to shun it. With every story of chemical leakage or spillage, more people refused to drink water downstream from chemical industries and pressured governments to spend millions of dollars on pipes to bring in water from farther upstream.

The media was caught in a classic dilemma of who to believe about the risk. Ever since the mercury incident of 1970, fish from a number of contaminated areas had been listed as unsafe for human consumption. We were told that chemicals in the environment were already harming wildlife, causing infertility, birth defects, abnormal behaviour such as abandonment of nests, cancer, and death. We were shown photos of cormorants born with crossed beaks and white suckers with skin tumours called papillomas. We were told that dead belugas that washed up on the shores of the St Lawrence had such high PCB levels that they qualified as hazardous waste. At the same time other scientists and government officials were saying the amount of pollution that reached people in drinking water was too low to cause problems. Sometimes there was confusion about what was a "safe" level of pollution. I remember staking out a closed meeting of the top Canadian and U.S. experts on the acceptable level of dioxin in fish that were to be eaten. The federal officials came out of the two-day session unanimous on the cutoff level for health advisories. A day later New York State announced a lower level. Why? Not because the scientists disagreed about the toxicity of the chemicals, but because New York officials assumed that people would eat more of the fatty tissue where most organochlorine contaminants concentrate.

Usually, the journalist has little contact with his or her readers, but fears about drinking water safety broke down the reluctance of people to call the media. I had been working on the water pollution issue when the phone rang one day. At the other end was not a scientist, bureaucrat, or activist, but a very nervous mother asking if it was safe to drink the water, or even to have children. All I could say was that the tap water was not perfect, but it did seem safe enough to drink. About the same time, I remember casually walking across the newsroom to drink at the water fountain. There was an audible sigh. I looked up in surprise and was told by one of my fellow reporters that they were relieved to see that I would still drink the water.

The drinking water story now appears less frequently on page one than it did in the mid 1980s. Studies of treated tap water from municipalities around the lakes and down the St Lawrence have found that contaminant levels are usually within acceptable guidelines set by federal, state, or provincial governments. Problems arise when there are spills close to drinking water intakes or leaking wastes near wells or when the drinking water treatment system fails to perform adequately. The greatest risk from chemical pollution is to people who eat a lot of food that has high levels of contaminants. Despite official assurances that tap water meets government guidelines, fear of pollution is driving people to buy tens of millions of dollars of bottled water and home water filters annually.

In the latter 1960s, my second newspaper job took me to the Detroit-Windsor area, an industrial centre long known as the world's auto capital. I soon began to understand that I was as likely to get pollution in the air I breathed as in the tap water. By the time I finished an eight-hour shift, I had to lean on the windshield washer button so I could see to drive home. A colleague wrote that breathing the air in that region was like smoking two packs of cigarettes a day. More than twenty years later, a 1990 IJC study of air pollution around Detroit, Windsor, Sarnia, and Port Huron put it in more scientific terms. It listed 125 air pollutants, including arsenic, chloroform, formaldehyde, and benzene – substances that can cause cancer and reproductive problems and harm the immune and nervous systems. The sources of air pollution included thousands of businesses, large and small, 1,700 incinerators, and hundreds of thousands of cars.

As your plane lifts off the runway of big city airports from O'Hare in Chicago to Dorval at Montreal, you rise through a brownish haze. After a couple of minutes the plane climbs through the top of the smog layer, and you realize that you are breathing in a chemical soup of nitrogen, petroleum, and other waste gases constantly exhaled by the machines that run our industrial society. People living in and downwind from cities have become accustomed to rarely seeing the crystal blue that lies above the smog. One of the pollutants that make skies so drab is the sulphate particles that cause acid rain. This corrosive fallout, formed when air pollution from industries, power plants, and cars mixes with water in the sky, will not turn the Great Lakes or St Lawrence River into vinegar: their limestone-based geology neutralizes the acid. The acidic rain and snow does, however, attack thousands of smaller lakes and rivers from the Adirondacks to the Canadian Shield, and from Minnesota to the Maritimes, gradually killing off populations of fish and creating "acid-dead" lakes.

Driving north into central Ontario's holiday region, I found cottage owners who feared that their summer retreats on inland lakes were turning into sterile pools of water. Farther north, the fishing lodge owners were incensed and militant, demanding government action to protect the environment that was the base of their business. I walked the shores of a small Muskoka lake with journalists from as far away as Chicago. They peered into the crystal clear water, and wondered how to explain to desk-bound editors that this beautiful lake was being harmed by pollution from as far away as power generating stations in the Ohio valley and smelters in Ontario. Finally, one journalist turned to a scientist acting as our guide, and demanded proof that this really was a dead lake. The scientist jabbed at a battery of boxes and white plastic tubes running into the lake, as if they were a life support system, and patiently explained how tests had shown increasing levels of acid in the water and fewer fish. Where were the dead fish, the evidence? asked another reporter. At the bottom of the lake, the scientist replied.

The government scientists had another way of convincing journalists that pollution from tall smokestacks was noxious. They loaded us into twin-engine aircraft and flew us north to the mining and smelting centre of Sudbury, where acid gases poured out of skyscraper tall smokestacks of copper and nickel smelters. As the aircraft did figure-of-eight turns through the plumes of sulphur dioxide gas, entire plane-loads of journalists turned green.

The air pollutants came from and harmed both countries. I spoke with people in Pennsylvania angry that acid rain was destroying their trout streams and their history. The corrosive fallout was literally erasing inscriptions from the monuments in Gettysburg National Park, the world's largest collection of outdoor statuary.

In the Beauce region south of Quebec City I stood with forest ecologist Gilles Gagnon in the midst of a dying maple bush. "Last year, there was a forest here," he said. Now, many trees were dead, and the survivors looked sick, with yellow and brown leaves, their edges curling and shrivelling. Many leaves were perforated with small holes. The cause of this forest dieback, which was found in a number of polluted regions of North America and Europe, has been hotly debated for years. Gagnon and many other foresters believed that acid rain was one of the culprits. So did farmer Michael Herman, in Quebec's Eastern Townships. I was travelling with a camera crew hunting for signs of acid rain damage for a film on water when we arrived at his sugar bush one cold, rainy morning in late winter. Walking with Herman, I could feel the anger seething in him as he pulled rotting pieces of bark from his trees and pointed to dead branches overhead. He was convinced that the cause was the invisible hand of acid air pollutants from faraway industries. It is a hand that falls unevenly on the environment below.



On the north shore of the St Lawrence, not that far away from Michael Herman's bush as the winds blow, we drove Route 138, "le Chemin du Roy," constructed more than 250 years ago when the land was the colony of New France. The road winds past sugar bushes that have been tapped for generations. Here we found farmers happily boiling the syrup in their long, shallow trays. Acid rain? They had hardly heard of the term and, no, their maple bushes were in fine health. But while acid precipitation does not usually kill vegetation outright, it adds one more stress to such natural pressures as drought, disease, and insects. Some scientists feel the effects of acid fallout are enough to push already weakened forests into decline.

The lakes do not have tides like the oceans, but lake levels can change dramatically over time. They fluctuate depending on the rain and snowfall. Over a period of years the same spot on a shoreline will be bone dry at the low point, and the water waist high at the high-level mark. My personal gauge of the lake levels is "The Rock" in my home town. It is large and blackish-brown. Whatever angles it once had were polished away by millennia of sand and waves. When it stands high above the water, I know the lake levels are low. When the waves lap over it, the lakes are up.

During a low-water period of the mid 1960s I was driven in an old army truck to what had been an island in Lake Huron. We splashed across a channel where men in aluminum boats had been trolling for bass in the evenings a couple of years before. Now it was almost part of the mainland. Within a year or two, water levels began to rise, and two decades later, after years of above-average rainfall, the lake levels were at their highest in 120 years. For me the high-water level was a minor inconvenience. I could no longer walk the length of my favourite beach, because the waves now buried the sand and lashed against the trees. One of my favourite lakeshore roads was washed out and piled high with ice floes during the winter. For many other people it was a very serious business. Around the Great Lakes hundreds of people were periodically forced to flee their homes when storm-whipped waves pounded across front yards and battered down their doors. Dozens of homes were destroyed. Some people gave up and moved away from the lakeshores, unwilling to joust with nature. Others called for governments to control the lakes by dredging the rivers that connect the lakes to carry away the water. Canada and the United States did intensive studies but decided against major new works. They concluded that, given the cost of the work, unpredictability of rainfall, and the length of time it would take to move the water out of the lakes, it was not feasible to control nature on this scale.



Just as a forester can tell not only the age of a tree but much about the region's history by examining tree rings, limnologists can look at the history of our waters. They examine the varves, those thin layers of clay and silt on lake and river bottoms that mark the deposits of each year. Tests of these annual layers of sediments, particularly in areas such as the mouth of the Niagara River, indicate that toxic chemical pollution rose with the explosive growth of industry starting around 1940. The deposits of arsenic, cadmium, lead, mercury, DDT, dioxins, furans, and other industrial chemicals and pesticides peaked during the 1960s and 1970s, then began to drop sharply.

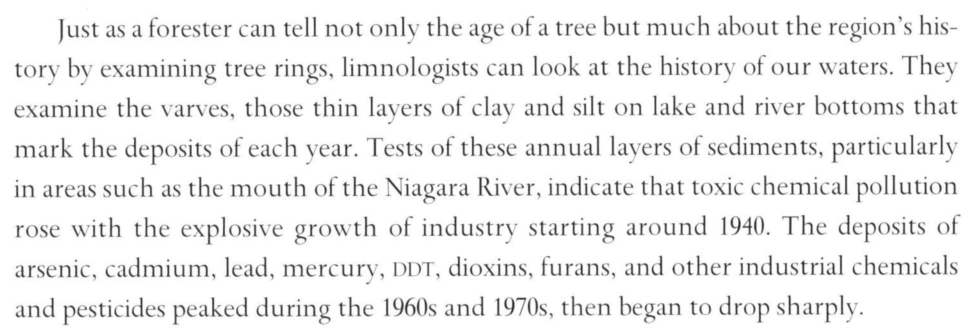

By the late 1960s public revulsion at the mess in our lakes and rivers was forcing political action. In 1972 the tide began to turn against uncontrolled pollution, when Canada and the United States signed a historic Great Lakes Water Quality Agreement. By 1978 the two nations were back at the table, signing a second and much broader

agreement that pledged to restore and maintain "the chemical, physical and biological integrity of the waters of the Great Lakes Basin Ecosystem." This pact marked a breakthrough in pollution control by calling for the virtual elimination of discharges of persistent toxic substances. In 1987 the agreement was expanded again to cover such diverse pollution sources as toxic chemical fallout from the air, leaking chemical waste dumps, and polluted runoff from farm fields, industry, and city streets.

Since the late 1960s, and particularly during the 1980s, there has been a sharp reduction in dumping of chemical wastes into the environment. There have also been bans and restrictions on the use of a number of toxic substances, including PCBs, DDT, and mercury. In the past twenty-five years, Canada and the United States have spent about $9 billion at state, provincial, and federal levels, and probably as much has been invested at the municipal level. Over two hundred communities in the Great Lakes and St Lawrence basin have made improvements to their sewage systems, including the installation of more than 150 new treatment plants. By law we have reduced phosphorus levels in laundry detergents. Industries have invested billions of dollars in pollution controls and cleanups. In a number of cases the investments have paid off not only ecologically but financially. They have reduced the amount of product lost to the environment, cut energy costs, and reduced the cost of waste disposal.

More than two decades of phosphorus controls have nursed the sick Lake Erie back to better if not perfect health. Its waters are much clearer than they have been for at least a generation, and the slimy green algae have retreated from the beaches. The Cuyahoga River is far from clean, but at least it does not catch fire, and the waters at the foot of Niagara Falls are no longer mounded high with polluted soapsuds or brown with sewage. In recent years I have begun to see the sooty black double-crested cormorants flying fast and low in front of my favourite island. Now that DDT has been controlled, it no longer builds up in their bodies to levels that cause the eggs to break in the nest. These fish-eaters have reclaimed their old nesting areas and are once again part of the ecosystem. The cormorants share the island with the traditional great blue herons and newcomers of the past few years, great egrets. A scientist I know well told me that the island, one of the important wildlife nesting sites on the lakes, has long been a barometer for toxic chemical levels. Those levels have been declining in most parts of the lakes.

Walking the urban shorelines is not always as pleasant as a hike along a wild beach. Chimneys and office towers, not trees, form the skyline. The urban waves often bring an ugly mix of flotsam ranging from discarded soft-drink bottles to old wrappers. The many streams that once laced our cities are now often invisible, except where they empty into a lake or river from the mouth of a drainage pipe. At first the streams were sources of drinking water, then open sewers. When the stench got too bad, they were buried and turned into storm sewers. Walking the banks of Black Creek in Toronto, I discussed the fate of urban waterways with a former environment minister, a zoology professor, a landscape architect, and a university student. What they had in common was a belief that the degraded waterways could be reclaimed. They believed that it might even be cheaper to clean up the water by such activities as planting trees on riverbanks than to build huge sewage treatment systems to purify what have become waste waters.

In Buffalo I walked with experts in urban renewal down to the waterfront along a newly paved and decorated street. The giant steel plants that had long defined the city had been closed for years, replaced by cheaper and sometimes cleaner mills in other states or other countries. Local politicians had decided they wanted the image of a clean city and were working hard at urban renewal. This is a sentiment echoed in Hamilton, Canada's steel city. Even in the mid 1980s officials regarded the pollution damage to Hamilton Harbour as so severe as to be as to be virtually incurable. Industry has turned off most of the pollution taps, however, and local citizens are working to reclaim what had been one of the most beautiful bays on the lakes. Now when you drive the Burlington Bay skyway and look over Hamilton, the skyline is no longer obscured with smoke and grit. The bay is slowly starting to heal.

One of the best ways to see Detroit's impressive skyline is to stand on the riverbank in Windsor. From that perspective the new office towers, built as part of the urban renewal projects of the 1970s, glisten in the sun. Detroit has also built a gigantic sewage treatment plant which has dramatically reduced the phosphorus pollution coming from the city into the Detroit River and Lake Erie. But like most treatment plants, it is having a harder time removing toxic chemicals.

An observation about how new and controversial ideas are adopted by society says they have to go through three stages. At first, the establishment reacts by saying the new idea is crazy and dangerous. If the idea is not squelched but finds a growing number of supporters, the establishment becomes quiet and watchful. Finally, if the idea gains widespread popular support, the establishment jumps on the bandwagon saying, "We were in favour of this idea all along."

So it has been with the attitude toward the environment of the public, government, and business over the past three decades. When scientists and environmentalists raised the warning flags against pollution in years past, companies often denied that the pollution was a real problem or said that only a few people were affected and that clean-ups were too costly. Governments generally adopted this viewpoint and saw their role as helping industry to function without hindrance. At times governments even passed laws to protect the right of companies to pollute.

The attitude has slowly pivoted 180 degrees. This turnaround began in the 1970s and grew through the 1980s. Some of it came about through the enlightened actions of a few government and business leaders, but they were in the minority. The great moving force was a change in public opinion. People lost their innocence about pollution, became disgusted by it, then fearful for their health and for the risks to the beauty of nature. This shift in public attitude was driven by the actions of a relatively small number of people: scientists and environmentalists, working alone or in small organizations. Now most companies realize their environmental behaviour is crucial to their long-term success. As one chemical-industry leader told a public meeting in the late 1980s, companies have public permission to operate in society. If the public becomes too angry with an industry, that permission to operate can effectively be revoked through a combination of regulations and boycotts of products. Companies are periodically reminded of this by demonstrations or consumer boycotts.

The discoveries of environmental pollution were generally made by scientists. Rachel Carson was one of the most important in this field, because her book *Silent Spring* awakened the public to the threats of careless use of chemicals. She and other scientists put their reputations, even careers, on the line to speak out about environmental risks. This was especially true for those working for governments that did not want environmental quality to become a major public issue.

It was the environmental movement that began to blossom in the 1960s that translated the scientific jargon into terms understandable by the average person. The environmentalists often added an ethical component to the issues, asking, "Is it right to dump this waste into the river, or to cut this forest?" At first they were small groups of friends and neighbours rallying to protest against a dump or the spraying of chemicals. Sometimes they were a group of university students carrying a coffin to the edge of a polluted river to symbolize its dying. In many cases, the "group" was one person who had motivated friends and neighbours. I watched one woman with a telephone and a book on chemistry begin a movement that stopped a multi-million dollar waste disposal centre.

In the early years environmental organizations were seen as nuisances by the establishment. As they became more effective, these groups began to be accepted as part of the political landscape – not welcomed by government and business, but recognized as a political force. In recent years they have come to be seen as a legitimate third force in making recommendations, even helping to shape government and corporate decisions about the future of our environment. Over the past couple of decades some of these small volunteer groups have evolved into highly organized research and lobbying groups. Now they are often consulted by government and business and asked to represent the "public" interest.

Yet despite the tens of thousands of stories about pollution, despite the billions spent on sewage treatment plants, and despite the tremendous change in attitudes toward the environment, our waters and our watersheds are far from clean, let alone healthy. Towns and cities discharge so much bacteria-laden sewage that the lake water they draw back in is not safe for drinking without disinfection. Many municipal beaches are too polluted for swimming, especially after rains overload treatment plants, washing raw sewage into the lakes and rivers.

The levels of toxic chemicals have dropped but not vanished. Bird species such as the gull, cormorant, and eagle no longer face mass poisoning, but they are not all healthy. Many young are still deformed. Even though laws in Canada and the United States have reduced the discharges of a number of toxic chemicals, they have not virtually eliminated them. Discharges continue to be allowed under permits, most of which are still based on the principle of dilution of waste discharges. While point sources such as industries and municipal sewage systems are under stricter controls over the past twenty years, there is now a need to control the so-called non-point sources such as urban streets, storm sewers, farm fields, and the atmosphere.



When you look at a satellite photo map of the Great Lakes–St Lawrence watershed, the blue of the waters and the green of the forests appear the same whether in Minnesota, Quebec, or New York. It is only when you spread out a political map on the table and see the many boundaries that you realize how we have fragmented the way we approach the natural world. Years of exploration, fur trading, wars, and political negotiation have carved up this ecosystem into two nations, two provinces, nine states, and hundreds of county, city, town, and regional governments. We know that political boundaries mean nothing to the flow of water, growth of trees, and migrations of fish and birds, or the movement of pollutants. Governments are starting to realize that it makes sense to follow watershed and ecosystem boundaries; but it is difficult to override centuries of political habits.

The most important attempt to foster an ecosystem identity over national self-interest has been the International Joint Commission. The IJC does not completely reflect the drainage basin, because its jurisdiction stops where the international boundary leaves the St Lawrence River upstream from Montreal. But Quebec is often represented at IJC meetings by an observer, and there is increasing discussion of the impact on pollutants from the Great Lakes on the St Lawrence River. About 40 percent of the contaminants in that river arrive from the Great Lakes.

It has been interesting to watch the IJC at work. Its members are political appointees, usually with little training in ecological issues. Nonetheless, a number of politicians appointed to this body have come out with strong environmental statements, after seeing the mountain of evidence about damage to the environment. Advisors to the IJC usually come from federal, state, and provincial government environmental agencies. These are the same people responsible for enforcing pollution controls. Yet when they have to speak to the IJC, I have heard some of these bureaucrats say, "I will put on my IJC hat." This means they will try to set aside the sometimes narrow interests of their governments and speak about what is good for water quality.

As a result of this approach, the IJC has emerged as a powerful and influential watchdog. Its reports on the forty-two toxic hotspots around the basin are a constant reminder that a lot remains to be done. The IJC is less active at times when governments shuffle members or leave positions unfilled, but it stands out as one important attempt to face the realities of environmental problems.

## Perspectives

In a half millennium of European exploration and settlement the St Lawrence River and the Great Lakes have been cast in many roles. First thought to be a western trade route to the Orient, they then became the explorers' pathway into the New World, with its own riches. Following closely behind were the priests, soldiers, fur traders, and finally the settlers who established new civilizations, shouldering aside those of the First Nations. Later the lakes and tributary rivers were harnessed to power grist and saw mills, carry freighters, cool molten steel, and make chemicals. They provided drinking water for a rapidly expanding society and washed away its wastes. Across this basin, forests have been cleared, wetlands drained, pits and quarries carved, fields turned into farms, and farms paved over for towns and cities. In the southern part of the watershed only a tiny fraction of the pre-pioneer landscape remains.

Settlement patterns have ebbed and flowed. The first wave of immigrants tried to farm almost every acre of land that could be fenced, including many areas too rocky or sandy to support even a modest lifestyle. Through the late nineteenth and early twentieth centuries less fertile lands were gradually abandoned, especially as the rich western prairies were put to the plow. The area where I grew up has been through a series of cycles as the boom times of forestry, fisheries, and agriculture in the last century were followed by declines. Many farms were reclaimed by wildflower meadows, poplar groves, birch stands, and cedars. Following animal trails through the dense bush as a youth, I would periodically stumble on the foundations of an abandoned homestead or the rotting remains of a cedar rail fence. To me this second-growth forest was like untouched wilderness. But in the past couple of decades a new wave of settlement has come, bringing more summer cottages, retirement subdivisions, strip malls, and factories. The marshes that once wet my hiking boots and the streams where I pursued trout and smelt are now drained by a growing network of ditches and storm sewers. The trackless swamps of my childhood are now criss-crossed with roads, and sand dunes have been bulldozed flat to make way for driveways and lawns. The deer, muskrat, grouse, and salamanders have been pushed back into ever-smaller fragments of habitat.

As more people move to the country, rural areas are transformed into suburbs, then towns and cities. The lifestyle involves the consumption of large quantities of water to maintain Kentucky-bluegrass lawns in regions where they would normally wither and die. It encourages the use of herbicides and insecticides to keep yards looking like carpets rather than natural meadows. Instead of using rain water on our gardens, we pipe it off our roofs into drains, overloading sewage-treatment plants to the point that during rainstorms raw wastes must be released to pollute the beaches. Then, we spend billions of tax dollars building bigger sewers, and millions of our own dollars on swimming pools filled with chlorinated water. Every summer when our lawns dry out, we pipe water back up from the lakes and rivers, thus creating the need for larger pumping, treatment, and delivery systems.

Merely living on this planet, eating, working, travelling and consuming resources, we have an environmental impact. We all leave a personal ecological footprint in terms of fields plowed, trees cut, holes dug in the ground, asphalt laid, and pollutants dispersed. Above a certain minimum level of consumption needed to survive, we have the choice to make our ecological footprint larger or smaller. Our consumption shapes our environment, and our personal values shape those consumption patterns. If we prize cars with big engines, and houses with big lots, we should expect to get more smog and pavement and less wilderness in return.

Until recent years most of us lived in ignorance of our environmental impacts or considered them less important than economic growth. Now virtually everyone is aware that we have serious environmental problems created not just by big business but by the actions of a whole society. The small amount of waste from an individual seems trivial, but when the drips and drops of pollution from each of tens of millions of people are added up, the effect is dramatic. Conversely, if we all reduce our individual impact, the common environment becomes cleaner. Behaviour is starting to change. A growing number of people routinely carry a recycling box to the curb, buy recycled paper, and reduce the amount of synthetic chemicals they spread on the landscape or pour down the drain.

Living and consuming in an ecologically sustainable manner – one that does not run down the environment – requires information and action. We can easily get information about the state of our ecosystems in free or inexpensive reports from governments, non-governmental groups, industries, and universities. There is plenty to read and see in newspapers, videos, and hundreds of books and magazines in bookstores and libraries. Computer networks and the Internet can deliver everything from satellite images of the planet to newsletters on environmental community action. In a democratic society, we have a responsibility to use that information in deciding how we vote and what we buy. Politicians will respond to pressure for laws that protect our environment. They will also take lack of pressure as a signal that this is not a priority, particularly in a time when they are looking for areas to reduce spending. Businesses flourish or vanish based on what the public thinks of their products and performance. Buying "green" products will encourage their production, while avoiding environmentally unfriendly goods will signal companies that they are not worth producing.

When European settlement began, the waters were pure enough to drink and teemed with fish and wildlife. We should not expect to roll back the ecological clock, but we *can* undo much more of the damage. It is up to all of us to decide what kind of environment we want. Do we want a landscape made of concrete? Should the waters be made clean enough just for boating – or for fishing, swimming, and drinking also? These are not new questions. All around the watershed, teams of people have been working for years to set clean-up goals for dozens of contaminated areas. These restoration plans represent the collected wisdom and vision of governments, industries, and citizens.

What would be the signs of a healthy ecosystem? Pollution levels would be so low that we could stop worrying about the safety of our food and drinking water. One visible symbol of success would be bald eagles soaring again above all the lakes and rivers. If these creatures at the top of the food chain are healthy, we will have an environment where persistent toxic substances no longer pose a threat. There would be more marshy areas, those cradles of life and purifiers of the waters. Our cities would still be active centres of living and working, but children could again enjoy their beaches without fear of getting sick. At the edges of the cities there would be farms, but they would no longer be draining herbicides, pesticides, soil, and animal wastes into the creeks and rivers. Among the farms would be woodlands big enough to support a diverse collection of plants and animals, including many species the pioneers would recognize. Upland and headwater regions would have large expanses of healthy and diverse forests. And throughout the entire Great Lakes and St Lawrence watershed, there would be significant lands and waters protected and left wild, where nature can just *be* – for its own sake.

Perhaps if we are determined enough to clean our environment, our lakes and rivers and the surrounding landscape will give us the sense of wonderment and well-being that so strongly touched the first visitors to this land.

The "Voices for the Watershed" project was conceived in 1994. In this epilogue, co-editor Bruce Litteljohn explains how it all began with a seemingly innocent photographic expedition aboard the RV Edwin Link. He concludes with a synopsis of issues facing the region, and a plea for each and every one of us to protect and restore the Great Lakes–St Lawrence Watershed.

Research Vessel *Edwin Link*. Photo: Frozenrope

# Epilogue

Bruce Litteljohn

The geographical and topical magnitude of this book is unusual. In fact, during the past five years as it took shape, the editors sometimes felt that the arrival of the millennium was small potatoes compared to bringing this work to completion. For while the book may not be as deep as Lake Superior (406 metres or 1,330 feet at its deepest) or the Saguenay River fiord (320 metres or 1050 feet) it is very broad in terms of the huge area covered and the variety of subjects tackled.

It began more narrowly.

In June 1994 a call came requesting that I go on board the Research Vessel *Edwin Link*, under the command of Captain James V. Seiler of Harbor Branch Oceanographic Institution. The ship, which carried the three-person submersible *Clelia* on her aft deck, had sailed north from Fort Pierce, Florida, to undertake a series of approximately sixty dives in the Great Lakes, the Saguenay River, and the St Lawrence. The unique vessel and submersible were in the region for several purposes based on several funding sources. The chief purpose of the voyage was sub-surface scientific research and observation sponsored by the National Undersea Research Center (NURC) at the University of Connecticut's Avery Point campus. The *Clelia* with its large acrylic port, hydraulic arm, video cameras, and lake-bed sampling technology was designed for these tasks. A second purpose was education. Peter Sheifele of NURC (an agency of the National Oceanic and Atmospheric Administration) worked with high-school students from both the United States and Canada, some of whom participated in submersible dives. This exciting experience was made possible through the Aquanaut

Program and sponsored by the Max and Victoria Dreyfus Foundation. For the students it opened new frontiers of knowledge. The third purpose was to focus media and public attention on the state of the St Lawrence and Great Lakes. As a photographer, I was invited to support this last function. However, other interests soon came into play.

For me photography had always been in large part an adjunct to volunteer work for a wide variety of conservation groups in both Canada and the United States. At the same time, I had been a director of the Wildlands League (a chapter of the Canadian Parks and Wilderness Society) for about twenty-five years, a teacher of environmental studies at Toronto's Upper Canada College, and director of the College's Norval Outdoor School on the Credit River (which flows into Lake Ontario) for fifteen years. At our outdoor school, students and the Ontario Ministry of Natural Resources had by 1994 planted more than 700,000 trees, largely as an exercise in forest and wildlife restoration and watershed protection. As well, I had paddled many watershed lakes and rivers and much of the Canadian shore of Lake Superior – an unforgettable and magnificent experience.

And so on board the R.V. *Edwin Link* it was impossible for me to function or think simply as a photographer. This was especially the case after meeting Dr Douglas S. Lee, then NURC's Freshwater Program director at the University of Connecticut. A highly experienced diver and an expert on both manned and unmanned submersible craft, Doug had been responsible for directing at-sea scientific operations for NURC, an agency of the National Oceanic and Atmospheric Administration, on Lake Baikal in the U.S.S.R., Lakes Victoria and Tanganyika in Africa, and Lakes Yellowstone and Champlain in the United States. On the Great Lakes he had often served as chief scientist aboard various research vessels. Watching him and his colleagues at work opened a window for me on marine science and the underwater world – in many places a sub-surface wilderness, in others a much manipulated and polluted environment.

It rapidly became apparent that the land-based research, conservation, protection, and restoration initiatives, with which I was familiar, had their aquatic or marine counterparts. Equally apparent was the fact that what goes on below the water's surface (often out of sight and out of mind) can have a tremendous impact on terrestrial life – and, of course, what we do on land has great (mostly negative) impact on marine ecosystems. In this respect, it may be very wise to recall ecologist Barry Commoner's warning, "Mother Nature always bats last."

My first lesson in how underwater research can illuminate our land-based activities: I first went aboard the *Edwin Link* at Oshawa Ontario near the Pickering Nuclear Power Station which is situated just east of Toronto. On several occasions my students and I had approached it from the landward side to tour the place, listen to the comforting official presentations, and then to ask the hard questions about disposal of radioactive materials, the enormous hidden cost of "decommissioning" worn-out reactors, the transport of radioactive fuel, and the frequency of leaks and accidents. Now the roles were reversed as, anchored about half a kilometre off-shore, I became very much the student with access to a large group of experts. On board were Dr Ken McMillan (geologist with McQuest Marine Research and Development, Hamilton, Ontario), Dr Robert Jacobi (Geologist, State University of New York, Buffalo), Dr A.A. Mohajer (professor of Earth Sciences and Environment, University of Toronto), Dr Joseph Wallach (Atomic Energy Control Board of Canada), Dr Roger Flood (State University of New York, Marine Sciences Research Center), Dr Gail McFall (geologist, Ontario Ministry of Northern Affairs and Mining), Dr Douglas Lee (NURC and University of Connecticut), and Steve Blasco, P. Eng. (Geological Survey of Canada, Bedford Institute of Oceanography, Nova Scotia).

I mention the many scientists involved to emphasize the concentration of expertise and the *cross-boundary* nature of the work. Here was an international group of scientists working together on an international lake in an international drainage basin. The lake and the watershed are works of nature; provincial, and state, and international boundaries are works of human aspirations and policies. Science and conservation work have to reach across those boundaries if our environmental problems are to be solved. Indeed, the old design adage that "form should follow function" has to be stood on its head so that function follows form. This belief was strengthened by the dives of the *Clelia* at the Pickering Nuclear Power Station, for I soon discovered that the purpose was to examine a potentially unstable fault line that runs almost under the reactor and then across the lake to New York State. The (land) form was there and the scientific function was appropriately one of international sharing. All to the good – but I was left with an uncomfortable question: Do we really want to position nuclear facilities practically on top of fault lines?



In the company of my friend and book designer V. John Lee, I next joined the *Edwin Link* at the picturesque Lake Erie fishing town of Port Dover. Here the purpose was to investigate a 130-year-old schooner that had gone down in fifty metres (165 feet) of water off Long Point – the "Graveyard of the Great Lakes" and a unique and valuable natural area protected, in part, by the Canadian Wildlife Service. Divers this time were surprised to find the wreck encrusted with alien zebra mussels, as most researchers had not believed that they could survive in waters that deep. Another voyage of discovery, it was one that had the important result of getting John Lee involved in discussion about a possible book.

There followed a long series of purely scientific dives in Lakes Michigan and Superior. These involved a wide variety of Great Lakes researchers, some of whom have contributed to this project. The next opportunity to rejoin the ship came at Whitefish Point, Michigan. The challenge here was to record events surrounding what might be called a "publicity dive" to the fabled *Edmund Fitzgerald*. Teachers from the Aquanaut Program took advantage of the opportunity to instruct high-school students from Sault Ste Marie about underwater submersible technology. On a beautiful "mill-pond" day we had teachers from Connecticut working with Canadian students who had embarked from a Michigan port to study almost in the middle of the world's greatest international lake. To further accentuate the cross-boundary nature of the work, a friend, biologist, and accomplished photographer, Scot Stewart from Marquette, Michigan, came aboard to share the photo making.

The fourth photographic shoot took me back east to whale country near the surging confluence of two mighty rivers – the Saguenay and the St Lawrence. Here, at Tadoussac – one of the earliest fur-trading sites in North America – it was easy to visualize Jacques Cartier's small wooden ship tacking upstream in 1535 on his remarkable voyage of discover. Now, centuries later on a very different ship equipped with every marvel of surface and sub-surface technology, we too set out on a much shorter voyage of discovery. Cartier, searching for a passage to the riches of Cathay, instead discovered, in the words of F. Scott Fitzgerald, "the fresh green breast of the New World." We, by contrast, were searching for a poisonous sediment in the deeps of the Saguenay fiord. Times change.

On Saturday, 16 July 1994, the *Edwin Link* hove to beneath the towering granite cliffs of Cap Eternité, a powerfully beautiful location. As kayakers passed quietly around the cape in what is now a protected marine park area, everything seemed pristine and almost unchanged since the days of pre-contact First Nations. But the massive cliff-face is matched by underwater cliffs plunging down more than a thousand feet to the dark bottom of the fiord. There, not all is pristine and clear. With major industries upstream, this is not surprising and has been known for many years. However, under the direction of Dr Charles Schafer of the Bedford Institute of Oceanography, the *Clelia* went down and took bottom cores to discover that toxic mercury pollution persists at levels higher than had been anticipated. As before, this scientific survey was shared with local students.

For a non-scientist and a teacher, life aboard the *Edwin Link* was revealing and exciting – entirely aside from photography. This was a floating, hands-on classroom with teachers deeply involved in fascinating and important research directed toward improvement of the environment – a far cry from my 1950s high-school science classes! And it reinforced a long-held conviction that those of us who are not scientists should really learn more of its language so that we can understand what is going on – *and* that scientists should more often attempt to present their findings in language understandable to the layperson.

New perspectives were provided during a helicopter flight downriver. As the *Link* steamed below the chopper, we could see white beluga whales swimming upstream, The importance of those thousands of tributaries that drain into the Great Lakes–St Lawrence became increasingly evident. At the same time, the bird's-eye view revealed vast stretches of forest seamed with streams flowing into the Saguenay, roads and patches of farm land, and in the distance the town of Tadoussac on the St Lawrence. From the air it was easy to see the relationship between land and water and to understand them as interdependent parts of watersheds.



On the long drive home to Toronto I began thinking that to make a book of value with something new to offer, a number of elements would have to be brought together. First, leading-edge scientific knowledge should be made available to the public in an understandable style. Second, an entire watershed approach would be necessary, including attention to land areas, tributary rivers, and streams as well as the Great Lakes and St Lawrence – the whole geographical area bounded by the immensely long height of land. Third, this watershed approach would in turn require a cross-boundary effort with writers and photographers from the United States and Canada, and an awareness that what is done – or not done – in Ohio or Michigan or New York State or Ontario or Quebec has an impact on the watershed, often far from the source. Fourth, it would be necessary to introduce a wide range of topics from many parts of the watershed including headwaters areas to make readers aware of both serious problems and important gains – the bad news and the good. Fifth, a main thrust of the book would be to raise public awareness of, and identification with, the watershed and specific parts of it. Sixth, the book would encourage scientific, ecological, conservation, and restoration activities within the watershed – partly by giving examples. Seventh, it would advocate broadscale citizen involvement – whether writing letters objecting to governmental cutbacks in environmental activity or planting trees along streambanks or joining non-governmental conservation groups – to improve the health of the watershed for all living things, from mink and moose to the kids on the block. That seemed enough to attempt!

By late August 1994, the Wildlands League chapter of the Canadian Parks and Wilderness Society – usually committed to narrower Ontario projects – had formally expressed support. The League too recognized that natural boundaries must sometimes override political ones where the protection of nature is concerned. Early in September a team made up of book designer V. John Lee, environmental consultant and writer Michael Keating, and myself travelled to the Atlantic coast (Avery Point) Campus of the University of Connecticut. There we met with Dr Douglas Lee, Dr Richard A. Cooper, and Ivar Babb of NURC. It was swiftly agreed that they shared the Wildlands League's enthusiasm for the project.

On 15 September, NURC at Avery Point wrote to Tim Gray, executive director of the League, to affirm a cooperative international project as well as committing very substantial seed money to push the work forward. We had come a long way from a bit of shipboard photography and casual conversation in the galley.

The load was soon lightened with the involvement of Gregor Gilpin Beck – a wildlife biologist with field experience on the Atlantic, Arctic, and Pacific coasts, and in the area of watershed ecology. For many years Gregor has led environmental education and community-based conservation projects for the bi-national Quebec-Labrador Foundation/Atlantic Center for the Environment, and for other organizations. He has published scientific papers and popular works, and has taught at Ryerson Polytechnic University and Seneca College in Toronto, and is now director of Conservation and Science for the Federation of Ontario Naturalists. With his background in science, and his knowledge of environmental education and the Great Lakes–St Lawrence bioregion, Gregor deeply involved himself in all aspects of the project and rapidly became a tower of strength.

With Doug Lee initially leading the way in terms of scientific expertise and contacts, the now international creative team began to meet and correspond regularly. Unfortunately, Doug later had to step back from the project due to other professional commitments. There was much to do for all members of the creative team – from fund-raising to thrashing out conceptual guidelines to discussing design, contacting and sometimes meeting prospective writers and photographers, and informing ourselves on many matters through research and field investigations.

As the work progressed over five long years, we learned a great deal from our expert contributors. We hope that we have been able to put this knowledge together in a way that interests our readers.

Now to the limitations. *Voices for the Watershed* is an *introduction,* with suggestions for further investigation. The drainage basin is vast. Whole books, and very big ones (such as Amy Demarest's 502-page *This Lake Alive!* [1997] on the Lake Champlain Basin) deal with sub-watersheds that we mention only in passing. Nor do we deal at length with the Finger Lakes of New York State or Lakes Nipigon and Nipissing in Ontario, or the Sandusky River in Ohio or the Moise and Gatineau in Quebec, or the Pigeon River that separates Minnesota from Ontario for part of its length, or the Brule in Wisconsin or the Genesee in New York State. To attempt it all would exhaust both us and you, and scare away any sensible publisher. We have therefore attempted to combine a number of overview essays with narrower ones that deal with problems common to thousands of specific areas around this great watershed, and solutions appropriate for them.

About 150 years ago James Russell Lowell, a distinguished American author and statesman, wrote that "In creating, the only hard thing's to begin; A grass blade's no easier to make than an oak" (*A Fable for Critics*, 1848). We have begun. Our writers, photographers, designer, and editors have put the root system of a tough, enduring, wide-branched oak in place. However, you, the readers, will form the trunk and branches. To accomplish this will mean active involvement in environmental matters.

We hope the tree will be sturdy enough to bear the weight of the environmental challenges ahead and that it will grow and proliferate faster than the problems.

How fast is fast enough? From the standpoint of water alone, much still remains unknown. Reporting earlier this year, Doug Cuthbert, an Environment Canada water issues expert, noted that "the amounts of water withdrawn, used and consumed in the Great Lakes–St Lawrence watershed are not well known or uniformly documented." Cuthbert also warned: "The quantities of water used for many purposes – agricultural, municipal, industrial – are increasing without limit."

Water withdrawals, whether from the lakes and rivers of the watershed or subterranean aquifers, have become a hot environmental and political topic. These withdrawals – for agriculture, industry, domestic use, and the surging bottled-water phenomenon – are largely the result of relentless population pressure, pollution of surface waters, and ill-informed profligacy. The central axis of the Great Lakes–St Lawrence watershed has been a population magnet since the years of early French exploration. In the last two centuries, the area has been transformed from an aboriginal hunting ground to the industrial heartland of North America. In 1993, Stanley Bolsenga and Charles Herdendorf noted in the *Lake Erie and Lake St. Clair Handbook* that "over 30 million people live along the Great Lakes, a hundred-fold increase in less than 200 years." The trend continues, including along the St Lawrence River.

Why should this pose a water problem, given the immense size of the central Lakes and the proliferation of lakes and rivers that surround them and help renew them? The key word is "renew." Only about 1 percent of Great Lakes water is consistently renewed by rainfall and rivers. The rest is known as "fossil water" – water that came from melting continental ice sheets about 12,000 years ago. This is not replaced. To interfere with this ancient bounty would be to mine the Lakes and reduce the volume of water flowing through them. The environmental and social costs of this could be devastating, bearing in mind that water is not a renewable resource. Its supply is fixed and finite.[1]

Water *quality* remains a major concern and one that is affected by our activities throughout the drainage basin. Academics and non-governmental organizations and remedial action groups struggle hard to improve the situation, and deserve our admiration and support. However, governments – and Ontario is a good example – have slashed budgets for environmental research, monitoring, and remediation, as well as relaxing some laws. This, no doubt, is what recently led Gary Gallon, a former senior environmental advisor to the Ontario government and now head of the Canadian Institute for Business and the Environment, to comment to the *Toronto Star* in September 1999 that "environmental protection has come to a standstill."[2] While perhaps overstated, his comment underlines the need for political wisdom and will, and for giving environmental protection much higher priority in the halls of government.

Aside from those who address or ignore the challenges, what is the real situation on the ground and in the waters? Well, there are more than 8,000 kilometres (5,000 miles) of Great Lakes shoreline in the United States, but as Marq de Villiers notes, "only 3 percent are fit for swimming, for supplying drinking water, or for supporting aquatic life." Warnings about fish consumption are in place on both sides of the international boundary. In 1998, the International Joint Commission reported that the lakes were cleaner than ever before. However, in the same report, the IJC noted overwhelming evidence of a "buildup of radioactive contaminants in the lakes from the nuclear power plant discharges." Marq de Villiers observes, "Each year, 50–100 million tons of hazardous waste is generated in the watershed for the lakes, 25 million tons of pesticides alone."[3] Were he to add in the St Lawrence portion of the drainage basin, the figures would be considerably higher.



What does our treatment of the watershed mean for the life within it? For some species it means fear of drinking or swimming in the water; for others, it means death. Anthony Ricciardi and Joseph Rasmussen, writing in 1999 in the journal *Conservation Biology,* raised a large red flag. Put simply, they suggest that ecological disaster in the form of lost biodiversity looms for North American lakes and rivers. Because of widespread environmental degradation, at least 123 freshwater species disappeared during the twentieth century. And they project that surviving species will die out at a rate of about 4 percent every ten years. In short, freshwater life – including fish, amphibians, mussels, and crayfish – are becoming extinct five times faster than land creatures. We

Facing page: Dragonfly. Photo: Scot Stewart

are more aware of the latter because they are more visible. But, while freshwater species may be out of sight and out of mind to most, they are essential to ecological processes and a healthy web of life of which we are a part. Ricciardi and Rasmussen place blame for the decimation of freshwater creatures squarely where it belongs: on us. The interacting stresses of human activities such as organic pollution, soil erosion, the draining of wetlands, toxic contamination, damming and dredging of waterways, along with the introduction or invasion of alien species such as the sea lamprey and zebra mussels, have led to a serious and sorry state of affairs.



There is much that can be done, and should be done

More to the point, there is much that *you* can do to help make a positive difference.

We hope for a fast-growing and tough-grained oak to bear the weight of the challenges ahead.

Please get involved by contacting any of the conservation and environmental organizations listed in the appendices.

# APPENDICES

# Conservation and Environmental Organizations



We hope that you have enjoyed *Voices for the Watershed*, and found it be an informative introduction to the numerous environmental challenges facing the Great Lakes–St Lawrence basin. The intent of this project, however, is to reach far beyond the level of providing information. We hope that the numerous "voices" that created this book motivate and inspire you – as individuals or as part of a group – to become actively involved in helping to protect and restore your local watershed.

The list of organizations that follows provides contact information for a few dozen major governmental and non-governmental groups involved in environmental issues in the Great Lakes-St Lawrence bioregion. All have websites, and many of these have links to related Internet sites. There are hundreds of additional groups that operate locally, regionally, or at the state and provincial level. All could use your assistance.

There are so many ways to help: you can clean up a stream, plant trees, pick up litter, monitor pollution, use environmentally friendly household cleaners, teach others about nature, or act as a private steward for wildlife and natural habitats. Please lend your support – and your voice – and become an active participant in the conservation of our waters, wildlands, and watersheds. Your help will make a difference.

**Wildlands League**
**(a Chapter of Canadian Parks and Wilderness Society)**
**380-401 Richmond Street W.**
**Toronto, Ontario M5V 3A8**
**(416) 971-9453 (WILD)**
**www.wildlandsleague.org**

American Rivers
1025 Vermont Avenue, NW
Suite 720
Washington, D.C. 20005
(202) 347-7550
www.amrivers.org

Bird Studies Canada
P.O. Box 160
Port Rowan, Ontario N0E 1M0
(519) 586-3531
www.bsc-eoc.org

Canada Centre for Inland Waters
Environment Canada
P.O. Box 5050
Burlington, Ontario L7R 4A6
(905) 336-4981
www.cciw.ca

Canadian Heritage River System
c/o Parks Canada
Ottawa, Ontario K1A 0M5
(819) 994-2913
www.parkscanada.pch.gc.ca/rivers

Canadian Nature Federation
606-1 Nicholas Street
Ottawa, Ontario K1N 7B7
613-562-3447
1-800-267-4088
www.cnf.ca

Canadian Parks and Wilderness Society (CPAWS)
National Office
506-880 Wellington Street
Ottawa, Ontario K1R 6K7
(613) 569-7226
1-800-333-WILD
www.cpaws.org

CPAWS, Ottawa Valley Chapter
506B-850 Wellington Street
Ottawa, Ontario K1R 6K7
(613) 232-7297
www.cpaws-ov.org

Canadian Wildlife Federation
2740 Queensview Drive
Ottawa, Ontario K2B 1A2
(613) 721-2286
1-800-563-WILD
www.cwf.fcf.org

Centre Saint Laurent/St. Lawrence Centre
Environment Canada
105 McGill Street, 7th Floor
Montreal, Quebec H2Y 2E7
(514) 283-5869
www.qc.doe.ca/csl

Commission for Environmental Cooperation
393 rue Saint Jacques ouest, bureau 200
Montreal, Quebec H2Y 1N9
(514) 350-4300
www.cec.org

Committee on the Status of Endangered Wildlife in Canada (COSEWIC)
c/o Canadian Wildlife Service
Environment Canada
Ottawa, Ontario K1A 0H3
(819) 997-4991
www.cosewic.gc.ca

Environment Canada
Inquiry Centre
351 St Joseph Boulevard
Hull, Quebec K1A 0H3
(819) 997-2800
www.ec.gc.ca

Federation of Ontario Naturalists
355 Lesmill Road
Don Mills, Ontario M3B 2W8
(416) 444-8419
1-800-440-2366
www.ontarionature.org

Fisheries and Oceans Canada
200 Kent Street, 13th Floor, Station 13228
Ottawa, Ontario K1A 0E6
(613) 993-1516
www.dfo-mpo.gc.ca

Great Lakes Commission
The Argus Building 2
400 Fourth Street
Ann Arbor, Michigan 48103-4816
(313) 665-9135
www.glc.org

Great Lakes United
Main Office
Buffalo State College, Cassty Hall
1300 Elmwood Avenue
Buffalo, New York 14222
(716) 886-0142
www.glu.org

Great Lakes United
Montreal Office
460 St Catherine ouest #805
Montreal, Quebec H3B 1A7
(514) 396-3333
www.glu.org

Greenpeace Canada
605-250 Dundas Street W.
Toronto, Ontario M5T 2Z5
(416) 597-8408.
1-800-320-7183
www.greenpeace.org

Greenpeace U.S.A.
1436 U Street, NW.
Washington, D.C. 20009
1-800-326-0959
www.greenpeace.org

International Joint Commission
Great Lakes Regional Office (Canada)
100 Ouelette Avenue, 8th Floor
Windsor, Ontario N9A 6T3
(519) 257-6703
www.ijc.org

International Joint Commission
Great Lakes Regional Office (U.S.A.)
P.O. Box 32869
Detroit, Michigan 48232
(519) 257-6715
www.ijc.org

National Audubon Society
Main Office
700 Broadway
New York, New York 10003
(212) 979 3000
www.audubon.org

National Sea Grant Office
1315 East-West Highway
SSMC-3, 11th Floor
Silver Spring, Maryland 20910
(301) 713-2448
www.nsgo.seagrant.org

National Undersea Research Center
National Oceanic and Atmospheric Administration
University of Connecticut at Avery Point
1084 Shennecossett Road
Groton, Connecticut 06340-6048
(860) 405-9121
www.nurc.uconn.edu

Natural Heritage Information Centre
Ontario Ministry of Natural Resources
Box 7000
Peterborough, Ontario K9J 4Y5
(705) 755-2167
www.mnr.gov.on.ca/MNR/nhic/nhic.html

Nature Conservancy of Canada
400-110 Eglinton Avenue W.
Toronto, Ontario M4R 1A3
(416) 932-3202
www.natureconservancy.ca

Nature Conservancy of Canada
(Ontario)
121 Wyndam Street North
Suite 202-204
Guelph, Ontario N1H 4E9
(519) 826-0068
www.natureconservancy.ca

Parks Canada
25 Eddy Street
Hull, Quebec K1A 0M5
1-888-773-8888
www.parkscanada.pch.gc.ca

Pollution Probe
12 Madison Avenue
Toronto, Ontario M5R 2S1
(416) 926-1907
www.pollutionprobe.org

Quebec-Labraïdor Foundation/
Atlantic Center for the Environment
55 South Main Street
Ipswich, Massachusetts 01938
(978) 356-0038
www.qlf.org

Quebec-Labrador Foundation Canada
(Atlantic Center for the Environment)
680-1253 McGill College Avenue
Montreal, Quebec H3B 2Y5
(514) 395-6020
www.qlf.org

River Network
National Office
1130- 520 SW 6th Avenue
Portland, Oregon 97204
(503) 241-3506
1-800-423-6747
www.rivernetwork.org

River Watch Network
153 State Street
Montpelier, Vermont 05602
(802) 223-3840
www.riverwatch.org

Sierra Club
85 Second Street, 2nd Floor
San Francisco, California 94105-3441
(415) 977-5500
www.sierraclub.org



Sierra Club of Canada
National Office
412-1 Nicholas Street
Ottawa, Ontario K1N 7B7
(613) 241-4611
www.sierraclub.ca

Sierra Legal Defence Fund
Ontario Office
300-106 Front Street East
Toronto, Ontario M5A 1E1
(416) 368-7533
www.sierralegal.org

St Lawrence River Institute of Environmental Sciences
1111 Montreal Road
Cornwall, Ontario K6H 1E1
(613) 936-6620
www.riverinstitute.com

The Nature Conservancy
Great Lakes Program
8 South Michigan Avenue
Suite 2301
Chicago, Illinois 60603
(312) 759-8017
www.tnc.org

Union québecoise pour la conservation de la nature
690, Grande-Allée Est, 4e étage
Quebec, Quebec G1R 2K5
(418) 648-2104
www.uqcn.qc.ca

United States Environmental Protection Agency
Region 5
77 W Jackson Boulevard
Chicago, Illinois 60604
(312) 353-2000
www.epa.gov

U.S. Fish and Wildlife Services
Department of the Interior
1849C Street NW
Washington, D.C. 20240
www.fws.gov

World Wildlife Fund Canada
410-245 Eglinton Avenue E.
Toronto, Ontario M4P 3J1
(416) 489-8800
1-800-267-2632
www.wwf.ca

World Wildlife Fund U.S.A.
1250 24th Street, NW.
P.O. Box 97180
Washington, D.C. 20077-7180
1-800-CALL-WWF
www.worldwildlife.org

Photo: Gregor G. Beck

# Selected References and Notes



**Prologue**

Balcer, Mary D., Nancy L. Korda, and Stanley I. Dodson. *Zooplankton of the Great Lakes: A Guide to the Identification and Ecology of the Common Crustacean Species.* Madison, WI, and London: University of Wisconsin Press 1984.

Clarke, Arthur H. *The Freshwater Molluscs of Canada.* Ottawa: National Museums of Canada 1981.

Colborn, Theodora E., Alex Davidson, Sharon N. Green, R.A. (Tony) Hodge, C. Ian Jackson, and Richard A. Liroff. *Great Lakes, Great Legacy*? Washington, DC: The Conservation Foundation; and Ottawa: The Institute for Research on Public Policy 1990.

Conn, David Bruce. "The Role for Research in the Recovery of the St Lawrence River." In *Sharing Knowledge, Linking Sciences: An International Conference on the St Lawrence Ecosystem,* Roger D. Needham and E. Nicholas Novakowski, eds. *Conference Proceedings.* Vol. 1. Ottawa: Institute for Research on Environment and Economy, University of Ottawa 1996.

Conn, David Bruce, and Colleen M. Quinn. "Ultrastructure of the Vitellogenic Egg Chambers of the Caddisfly, *Brachycentrus incanus* (Insecta: Trichoptera)." *Invertebrate Biology* 114 (1995): 334–43.

Geis, James W., ed. *Preliminary Report: Biological Characteristics of the St Lawrence River.* Syracuse, NY: State University College of Environmental Science and Forestry 1977.

Lanken, Dane. "Montreal Starts Cleaning up the St Lawrence." *Canadian Geographic* 110, no. 3 (1990): 28–35.

Merritt, Richard W., and Kenneth W. Cummins, eds. *An Introduction to the Aquatic Insects of North America.* 2d ed. Dubuque, IA: Kendall/Hunt Publishing 1984.

Peckarsky, Barbara L., Pierre R. Fraissinet, Marjory A. Penton, and Don J. Conklin, Jr. *Freshwater Macroinvertebrates of Northeastern North America.* New York and London: Cornell University Press 1990.

**Introduction**

Ashworth, William. *The Late Great Lakes: An Environmental History.* Toronto: Collins 1986.

Botts, Lee, and Bruce Krushelniki. *The Great Lakes: An Environmental Atlas and Resource Book.* 3d ed. Ottawa: Government of Canada and United States Environmental Protection Agency 1995.

Bouchard, Hélène, and Pascal Millet. *The St Lawrence River: Diversified Environments.* Montreal: St Lawrence Centre, Environment Canada 1993.

Colburn, T., J.P. Dumanoski, and J.P. Myers. *Our Stolen Future.* New York: Dutton 1996.

Colborn, Theodora E., Alex Davidson, Sharon N. Green, R.A. (Tony) Hodge, C. Ian Jackson, and Richard A. Liroff. *Great Lakes, Great Legacy?* Washington, DC: The

Conservation Foundation; and Ottawa: The Institute for Research on Public Policy 1990.

Federation of Ontario Naturalists. "The Great Lakes" (special issue). *Seasons: The Nature and Outdoors Magazine* 27, no. 3 (1987).

Fitzharris, Tim, and John Livingston. *Canada: A Natural History*. Markham, ON: Viking Studio Books 1988.

Government of Canada. *The State of Canada's Environment*. Ottawa: Government of Canada Publications 1996.

Hummel, Monte, ed. *Endangered Spaces: The Future for Canada's Wilderness*. Toronto: Key Porter 1989.

– *Protecting Canada's Endangered Spaces: An Owner's Manual*. Toronto: Key Porter 1995.

Keating, Michael. *To the Last Drop: Canada and the World's Water Crisis*. Toronto: Macmillan of Canada 1986.

Labatt, Lori, and Bruce Litteljohn, eds. *Islands of Hope: Ontario's Parks and Wilderness*. Willowdale: Firefly 1992.

Theberge, John B., ed. *Legacy: The Natural History of Ontario*. Toronto: McClelland & Stewart 1990.

Université Laval, Department of Geography. *Environmental Atlas of the St Lawrence*. Ottawa: Environment Canada 1992.

Weller, Phil. *Fresh Water Seas: Saving the Great Lakes*. Toronto: Between the Lines 1990.

### Highland Headwaters

Notes

1 "Watershed" or "drainage basin" refers to all the lands and waterways draining into the Great Lakes and St Lawrence River. This huge area is bounded by the "height of land" dividing our watershed from others such as the Ohio-Mississippi River drainage basin, or that of James and Hudson Bay or the Hudson and Mohawk River. The height of land tilts rivers in different directions to find different outlets to the seas.

2 Archibald Lampman, "Temagami."

3 John Elder in "Conservation at the Edge of Wilderness," *Wild Earth* 8, no. 4 (Winter 1998-99): 30-4, notes that the six Wilderness Areas within the Green Mountain National Forest, while very important to the dramatic resurgence of wildlife, are third-growth forest.

4 Cited in Bill McKibben, "An Explosion of Green," *Atlantic Monthly* 275, no. 4 (April 1995): 61-83. See also McKibben's *Hope, Human and Wild* (Boston: Little, Brown, 1995), 12-26.

5 This paragraph draws heavily on Paul Medeiros's "A Proposal for an Adirondack Primeval," Ceneozoic Society, *Wild Earth* (special issue): *The Wildlands Project* (1992): 32-41.

6 Personal communication, 18 November 1999.

7 Tim Tiner, "Opening the Flood Gates," *Seasons* 38, no. 4 (Winter 1998): 20.

8 Ibid., 23.

9 David M. Bolling, *How to Save a River* (Washington, DC: River Network, Island Press, 1994), 201-3.

10 J.B. Mansfield, *History of the Great Lakes* (J.H. Bears & Co., 1899).

11 Paul Shepard, "Romancing the Potato," from Coming *Home to the Pleistocene*, cited in *Wild Earth* 8, no. 3 (Fall 1998): 38.

12 Brad Cundiff, "Promised Lands," *Seasons* 39, no. 2 (1999): 23.

13 Reed Noss, *Maintaining Ecological Integrity in Representative Networks* (Toronto: World Wildlife Fund Canada, 1995).

14 Gerald Killan, *Protected Places, A History of Ontario's Provincial Parks System* (Toronto: Dundurn Press, 1993), 136. Killan's work sets the standard for parks system history worldwide.

15 Lampman, "Temagami."

16 The history of native, conservation, recreational, and industrial aspirations in Temagami is told with great

authority in Bruce Hodgins and Jamie Benidickson's *The Temagami Experience* (Toronto: University of Toronto Press, 1989). See also Killan, *Protected Places*, 372-3.

17 Cited in Hodgins and Benidickson, *The Temagami Experience*, 294.

18 *Toronto Star*, 9 August 1999, A13. Figures re areas of parks and conservation reserves are from *Ontario's Living Legacy, Land Use Strategy* (Queen's Printer for Ontario, July 1999).

19 Nathalie Zinger, "Quebec," in Monte Hummel (ed.), *Protecting Canada's Endangered Spaces*, 102-3.

20 World Wildlife Fund Canada, *Guardian of the Canadian Wilderness* (1999), 4.

21 John Washington et al., *Focus on Canada* (Toronto: McGraw-Hill Ryerson, 1978), 218-20.

22 Robert Bourassa, *Power from the North* (Scarborough, ON: Prentice-Hall, 1985), 4.

23 Quebec conservation groups are trying to make up for lost time. Environmental pressure groups did not exist until the late 1970s and then only in Montreal. See Harvey Mead, "Quebec's Natural Heritage," in Hummel (ed.), *Endangered Spaces*, 152-64.

*Maps*

Partnership for Public Lands. Candidate Protected Areas, Great Lakes–St Lawrence Region. Toronto: World Wildlife Fund Canada, Federation of Ontario Naturalists, Wildlands League 1998.

World Wildlife Fund Canada and Union quebecoise pour la conservation de la nature. *Les milieux naturels du Quebec méridional.* Toronto and Montreal: 1998.

*Books and Periodicals*

Ashworth, William. *The Late Great Lakes: An Environmental History.* Toronto: Collins 1986.

Bolling, David M. *How to Save a River.* Washington, DC: River Network, Island Press 1994.

Bouchard, Helene, and Pascal Millet. *The St. Lawrence River: Diversified Environments.* Montreal: St. Lawrence Centre, Environment Canada 1993.

Bourassa, Robert. *Power from the North.* Scarborough, ON: Prentice-Hall 1985.

Cundiff, Brad. "Promised Lands." *Seasons* 39, no. 2 (1999).

Elder, John."Conversation at the Edge of Wilderness." *Wild Earth* 8, no. 4 (Winter 1998-99).

Hodgins, Bruce, and Jamie Benidickson. *The Temagami Experience.* Toronto: University of Toronto Press 1989.

Hummel, Monte, ed. *Protecting Canada's Endangered Spaces.* Toronto: Key Porter 1995.

Killan, Gerald. *Protected Places: A History of Ontario's Provincial Parks System.* Toronto: Dundurn 1993.

Klees, Emerson C. *Persons, Places, and Things in the Finger Lakes Region.* Rochester: Friends of the Finger Lakes Publishing 1993.

Mansfield, J.B. *History of the Great Lakes.* J.H. Bears & Co. 1899.

McCarthy, John, and Linda Bishop McCarthy. *The Finger Lakes Revisited.* 2nd ed. Skaneateles: Finger Lakes Photography 1998.

McKibben, Bill. "An Explosion of Green." *Atlantic Monthly* 275, no. 4 (April 1995).

— *Hope, Human and Wild.* Boston: Little, Brown 1995.

Mead, Harvey. "Quebec's Natural Heritage." In *Endangered Spaces*, ed. Monte Hummel.

Medeiros, Paul. "A Proposal for an Adirondack Primeval." *Wild Earth* (special issue): *The Wildlands Project* (1992).

Noss, Reed. *Maintaining Ecological Integrity in Representative Networks.* Toronto: World Wildlife Fund Canada 1995.

*Ontario's Living Legacy, Land Use Strategy*. Queen's Printer for Ontario, July 1999.

Tiner, Tim. "Opening the Flood Gates." *Seasons* 38, no. 4 (winter 1998).

United States Environmental Protection Agency and Government of Canada. *The Great Lakes: An Environmental Atlas and Resource Book.* 3rd ed. 1995.

Washington, John, et al. *Focus on Canada.* Toronto: McGraw-Hill Ryerson 1978.

World Wildlife Fund Canada. *Guardian of the Canadian Wilderness.* World Wildlife Fund 1999.

### Over the Gunnels in Algonquin

Miller, Jeff. *Rambling through Algonquin Park: Paintings, Sketches, Thoughts by Jeff Miller.* Huntsville, ON: Limberlost Studio 1990.

Reynolds, William, and Ted Dyke. *Algonquin.* Toronto: Oxford University Press 1983.

Saunders, Audrey. *Algonquin Story.* Reprint, Toronto: Department of Lands and Forests 1963.

Seymour, Kevin, ed. *Algonquin Park at One Hundred: A Question of Survival.* Toronto: Wildlands League 1993.

Strickland, Dan. "Algonquin Park at One Hundred Years." In *Islands of Hope: Ontario's Parks and Wilderness,* edited by Lori Labatt and Bruce Litteljohn, 126–32. Willowdale, ON: Firefly 1992.

Theberge, John B., ed. *Legacy: The Natural History of Ontario.* Toronto: McClelland & Stewart 1990.

Troyer, Warner. "Algonquin Park – Some "Freeze Frames." In *Islands of Hope: Ontario's Parks and Wilderness,* 132–8.

### Getting our Act Together:

Website for l'Union quebecoise pour la conservation de la nature (including homepage for 'Scabric': http://eco-route.uqcn.qc.ca/frq

### Fouling up on the Farm

Faeth, Paul, Robert Repetto, Kim Kroll, QiDai, and Glenn Helmers. *Paying the Farm Bill: U.S. Agricultural Policy and the Transition to Sustainable Agriculture.* World Resource Institute 1991.

International Reference Group on Great Lakes Pollution from Land Use Activities (PLUARG). *Bibliography of PLUARG reports.* Windsor, ON.: International Joint Commission, Great Lakes Regional Office 1979.

Leopold, Aldo. *A Sand County Almanac: With Other Essays on Conservation from Round River.* New York: Oxford University Press 1966.

Novotny, Vladimir Novotny, and Harvey Olem. *Water Quality: Prevention, Identification, and Management of Diffuse Pollution.* New York: Van Nostrand Reinhold 1994.

Nonpoint Source Control Task Force of the Water Quality Board of the International Joint Commission. *Nonpoint Source Pollution Abatement in the Great Lakes Basin: An Overview of Post-pluarg Developments: Report to the Great Lakes Water Quality Board* (IJC report). Windsor, ON, and Detroit, MI: International Joint Commission, Great Lakes Regional Office [1983?].

### Wetland Lost – Farmland Claimed

Hawke, Stephen. *Wetlands.* Toronto: Stoddart 1994.

Mitchell, John G. "Our Disappearing Wetlands." *National Geographic* 182, no. 4 (1992): 3–45.

Mitsch, William J., and James G. Gosselin. *Wetlands.* 2d ed. New York: Van Nostrand Reinhold 1993.

National Wetland Working Group. *Wetlands of Canada, Ecological Land Class Services 24.* Montreal: Sustainable Development Branch, Environment Canada, and Polyscience Publications n.d.

Niering, William A. *Wetlands of North America.* Charlottesvilla, VA: Thomasson-Grant 1991.

Rezendes, Paul, and Paulette Roy. *Wetlands: The Web of Life.* San Francisco: Sierra Club Books 1996.

Scanlon, Kevin. "Life in the Muck: Toil and Trouble in Ontario's Holland Marsh." *Equinox* 20, no. 10 (1988): 68–81.

Van Patter, Mark, and Stewart Hilts. *Some Important Wetlands of Ontario South of the Precambrian Shield.* Toronto: Federation of Ontario Naturalists Report 1985.

**Wild Rice, Midges, and Cranes**

Bolsenga, Stanley, and Charles Herdendorf. *Lake Erie and Lake St Clair Handbook.* Detroit: Wayne State University Press 1993.

Crum, Howard. *A Focus on Peatlands and Peat Mosses.* Ann Arbor: University of Michigan Press 1988.

Cwiekiel, Wilfred. *Living with Michigan's Wetlands.* Conway, MI: Tip of the Mitt Watershed Council 1996.

Eastman, John. *The Book of Swamp and Bog Trees, Shrubs, and Wildflowers of Eastern Freshwater Wetlands.* Mechanicsburg, PA: Stackpole Press 1995.

Hawke, Stephen. *Wetlands.* Toronto: Stoddart 1994.

Mitchell, John G. "Our Disappearing Wetlands." *National Geographic* 182, no. 4 (1992): 3–45.

Mitsch, William J., and James G. Gosselin. *Wetlands.* 2d ed. New York: Van Nostrand Reinhold 1993.

National Wetland Working Group. *Wetlands of Canada, Ecological Land Class Services 24.* Montreal: Sustainable Development Branch, Environment Canada, and Polyscience Publications n.d.

Niering, William A. *Wetlands of North America.* Charlottesvilla, VA: Thomasson-Grant 1991.

Rezendes, Paul and Paulette Roy. *Wetlands: The Web of Life.* San Francisco: Sierra Club Books 1996.

Valpy, Michael. "Three Award-Winning Aboriginal Communities." *Globe and Mail,* 26 September 1995.

Williams, Michael C. Assistant Director, Nin. Da. Waab. Jig Heritage Centre, Walpole Island First Nation. Personal communication, 3 November 1999.

**Covering the Acid Rain Story**

Basil, John M. *Acid Rain: Its Causes and Its Effects on Inland Waters.* Toronto: Oxford Clarendon Press 1992.

d'Ayers, Harvey, Jenny Hager, and Charles E. Little. *An Appalachian Tragedy: Air Pollution and Tree Death in the Highland Forest of Eastern North America.* San Francisco: Sierra Club Books 1998.

Forster, Bruce A. *The Acid Rain Debate: Science and Special Interests in Policy Formation.* Ames, IA: Iowa State University Press 1993.

Minister of Supply and Services Canada. *The State of Canada's Environment.* Ottawa: Government of Canada 1996.

Mohnen, Volker A. "The Challenge of Acid Rain." *Scientific American* 259, no. 2 (1988): 30–8.

Pawlick, Thomas. *A Killing Rain: The Global Threat of Acid Precipitation.* San Francisco: Sierra Club Books 1986.

**The Media Agenda**

Keating, Michael. *Covering the Environment: A Handbook on Environmental Journalism.* Ottawa: National Round Table on the Environment and the Economy 1993.

**Reflections of a Paddle Pusher**

Mason, Bill. *Path of the Paddle: An Illustrated Guide to the Art of Canoeing.* Toronto: Key Porter 1984.

– *Song of the Paddle: An Illustrated Guide to Wilderness Camping.* Toronto: Key Porter 1988.

Raffan, James. *Summer North of Sixty: By Paddle and Portage across the Barren Lands.* Toronto: Key Porter 1990.

– *Wild Waters: Canoeing Canada's Wilderness Rivers.* Toronto: Key Porter 1986.

Ross, Alec. *Coke Stop in Emo: Adventures of a Long-Distance Paddler.* Toronto: Key Porter 1995.

**Poisons up the Food Chain**

Colborn, T., D. Dumanoski, and J.P. Meyers. *Our Stolen Future: Are We Threatening Our Fertility, Intelligence and Survival: A Scientific Detective Story*. New York: Dutton 1996.

Fox, G.A., D.V. Weseloh, T.J. Kubiak, and T.C. Erdman. "Reproductive Outcomes in Colonial Fish-Eating Birds: A Biomarker for Developmental Toxicants in Great Lakes Food Chains, 1: Historical and Ecotoxicological Perspectives." *Journal of Great Lakes Research* 17 (1991):153–7.

Moriarty, F. *Ecotoxicology: The Study of Pollutants in Ecosystems*. 2d ed. London and New York: Academic Press 1983.

Nriagu, J.O., and M.S. Simmons. *Toxic Contaminants in the Great Lakes*. New York: J. Wiley & Sons 1984.

Poole, A. *Ospreys: A Natural and Unnatural History*. Cambridge: Cambridge University Press 1989.

### Rescuing Native Fish

Edsall, Thomas A., and Gregory W. Kennedy. "Availability of Lake Trout Reproductive Habitat in the Great Lakes." *Journal of Great Lakes Research* 21 (supplement 1, 1995): 290–301.

Scott, W.B., and E.J. Crossman. *Freshwater Fishes of Canada*. Fisheries and Marine Science Bulletin 184. Ottawa: Ministry of the Environment 1974.

### Making the Lakes Great

Canada Centre for Inland Waters, Great Lakes Information Management Resourse, Remedial Action Plan Website: www.cciw.ca/glimr/raps .

Hartig, John H., and Michael A. Zarull, eds. *Under RAPS: Toward Grassroots Ecological Democracy in the Great Lakes Basin*. Ann Arbour, MI: University of Michigan Press 1992.

Hill Mackenzie, Susan. *Integrated Resource Planning and Management*. Washington: Island Press 1996.

International Joint Commission. *Great Lakes Water Quality Agreement*. 1987 Amendment to the 1978 Protocol. Windsor, ON, and Detroit, MI: International Joint Commission 1987.

Krantzberg, Gail. "After Ten Years of Effort Are RAPs Making a Difference?" *Journal of Great Lakes Research* 24, no. 3 (1998): 485.

– "Incremental Progress in Restoring Beneficial Uses at the Canadian Areas of Concern." Toronto: Ontario Ministry of Environment 1998 (also available at www.cciw.ca/glimr/data/analysis-10-years/intro.html).

Manno, J., et al. "Commentary on the Status of Remedial Action Plans." *Journal of Great Lakes Research* 23, no. 2 (1997): 212–36.

### From Apathy to Action

Ashworth, William. *The Late Great Lakes: An Environmental History*. Toronto: Collins 1986.

Botts, Lee, and Bruce Krushelniki. *The Great Lakes: An Environmental Atlas and Resource Book*. 3d ed. Ottawa: Government of Canada and United States Environmental Protection Agency 1995.

Colborn, Theodora E., Alex Davidson, Sharon N. Green, R.A. (Tony) Hodge, C. Ian Jackson, and Richard A. Liroff. *Great Lakes, Great Legacy?* Washington, DC: The Conservation Foundation; and Ottawa: The Institute for Research on Public Policy 1990.

Federation of Ontario Naturalists. "The Great Lakes" (special issue). *Seasons: The Nature and Outdoors Magazine* 27, no. 3 (1987).

Keating, Michael. *To the Last Drop: Canada and the World's Water Crisis*. Toronto: Macmillan of Canada 1986.

Kehoe, Terence. *Cleaning up the Great Lakes: From Cooperation to Confrontation*. DeKalb, IL: North Illinois University Press 1997.

Weller, Phil. *Fresh Water Seas: Saving the Great Lakes*. Toronto: Between the Lines 1990.

**The Crush of Megalopolis**

City of Toronto. "How to Solve the CSO Problem." *Safe Sewage News* 7. Toronto: Public Committee for Safe Sewage Treatment in Metropolitan Toronto 1995.

Hartig, John. H., and Michael A. Zarull, eds. *Under RAPs: Toward Grassroots Ecological Democracy in the Great Lakes Basin.* Ann Arbour, MI: University of Michigan Press 1992.

International Joint Commission. *Seventh Biennial Report on Great Lakes Water Quality.* Windsor, ON, and Detroit, MI: International Joint Commission 1993.

Minister of Supply and Services Canada. *The State of Canada's Environment.* Ottawa: Government of Canada 1996.

Niagara River Action Plan. *Niagara River Remedial Action Plan: Stage 2 Report.* Remedial Action Plan 1995.

Rodgers, G.K., Technical Team and Stakeholders. *Hamilton Harbour Remedial Action Plan, Stage One Report.* Hamilton: Governments of Canada and Ontario 1989.

– *Hamilton Harbour Remedial Action Plan, Stage Two, Main Report.* Hamilton: Governments of Canada and Ontario 1992.

Royal Commission on the Future of the Toronto Waterfront. *Regeneration: Toronto's Waterfront and the Sustainable City.* Toronto: Queen's Printer of Ontario 1992.

"A Sampling of Water Quality Facts." *Clean Water,* Fact Sheet no. 3. Ottawa: Environment Canada 1989.

"Toxics from Sewage Treatment Plants." *Eagle's Eye.* Toronto: World Wildlife Fund Canada (1995): 1–11.

"What Is Combined Sewer Overflow?" *Clean Water News.* Toronto: Metro Toronto and Region Remedial Action Plan 1994.

**Stories from a Big City Stream**

Don Watershed Regeneration Council. *Turning the Corner: The Don Watershed Report Card.* Toronto: Metropolitan Toronto and Region Conservation Authority 1997.

Don Watershed Task Force. *Forty Steps to a New Don: The Report for the Don Watershed Task Force.* Toronto: Metropolitan Toronto and Region Conservation Authority 1994.

Grady, Wayne. *Toronto the Wild: Field Notes of an Urban Naturalist.* Toronto: Macfarlane Walter & Ross 1995.

Gregory, Dan, and Roderick MacKenzie. *Toronto's Backyard: A Guide to Selected Nature Walks.* Vancouver: Douglas & McIntyre 1986.

Guthrie, Ann. *Don Valley Legacy: A Pioneer History.* Erin: Boston Mills Press 1986.

Hough Woodland Naylor Dance Ltd. and Gore and Storrie Ltd. *Restoring Natural Habitats: A Manual for Habitat Restoration in the Greater Toronto Bioregion.* Toronto: Waterfront Regeneration Trust 1995.

Ivy, Bill. *A Little Wilderness: The Natural History of Toronto.* Toronto: Oxford University Press 1983.

Sauriol, Charles. *Pioneers of the Don.* East York, ON: Charles Sauriol CM 1995.

– *Remembering the Don.* Scarborough: Consolidated Amethyst Communications 1981.

Simcoe, Elizabeth. *The Diary of Mrs. John Graves Simcoe: The Wife of the First Lieutenant-Governor of the Province of Upper Canada, 1792–6.* With notes and a biography by J. Ross Robertson. Toronto: Ontario Publishing Company 1934.

Scadding, Henry. *Toronto of Old.* Edited by Frederick H. Armstrong. 1873. Reprint, Toronto and Oxford: Dundurn Press 1987.

Seton, Ernest T. *Wild Animals I Have Known.* New York: Charles Scribner's Sons 1942.

Task Force to Bring Back the Don. *Bringing Back the Don.* Toronto: The Task Force to Bring Back the Don River 1991.

Theberge, John B., ed. *Legacy: The Natural History of Ontario.* Toronto: McClelland & Stewart 1990.

### Decline and Recovery

Rodgers, G.K., Technical Team and Stakeholders. *Hamilton Harbour Remedial Action Plan, Stage One Report.* Hamilton: Governments of Canada and Ontario 1989.

– *Hamilton Harbour Remedial Action Plan, Stage Two, Main Report.* Hamilton: Governments of Canada and Ontario 1992.

### Ecosystem in Peril

Bouchard, Hélène, and Pascal Millet. *The St Lawrence River: Diversified Environments.* Montreal: St Lawrence Centre, Environment Canada 1993.

Environment Canada, Quebec Region, St Lawrence Centre. *State of the Environment Report on the St Lawrence River.* Vols. 1 and 2. Montreal: Éditions MultiMondes, Environment Canada, St Lawrence Centre 1996.

Environnement Canada, Région du Québec, Centre Saint-Laurent. *Rapport-synthèse sur l'état du Saint-Laurent.* Vols. 1 and 2. Montreal: Éditions MultiMondes, Environnement Canada, Centre Saint-Laurent 1996.

Lanken, Dane. "Montreal Starts Cleaning up the St Lawrence." *Canadian Geographic* 110, no. 3 (1990): 28–35.

Ministère de l'Environnement du Quebec. *L'environnement au Quebec. Un premier bilan rapport synthèse.* Quebec, QC 1988.

Ministère de l'Environnement et de la Faune. *Liste de la faune vertébrée du Quebec.* Quebec, QC 1995.

Minister of Supply and Services Canada. *The State of Canada's Environment.* Ottawa: Government of Canada 1996.

Scott, W.B., and E.J. Crossman. *Les poissons d'eau douce du Canada.* (Freshwater fishes of Canada). Service des pêches et des sciences de la mer, bulletin 184. Ottawa: Ministère de l'Environnement du Canada, 1974.

Université Laval, Department of Geography. *Environmental Atlas of the St Lawrence.* Ottawa: Environment Canada 1992.

### Environmental Sleuths

Environment Canada, Quebec Region, St Lawrence Centre. *State of the Environment Report on the St Lawrence River.* Vols. 1 and 2. Montreal: Éditions MultiMondes, Environment Canada, St Lawrence Centre 1996.

Environnement Canada, Région du Québec, Centre Saint-Laurent. *Rapport-synthèse sur l'état du Saint-Laurent.* Vols. 1 and 2. Montreal: Éditions MultiMondes, Environnement Canada, Centre Saint-Laurent 1996.

Minister of Supply and Services Canada. *The State of Canada's Environment.* Ottawa: Government of Canada 1996.

Moriarty, F. *Ecotoxicology: The Study of Pollutants in Ecosystems.* 2d ed. London and New York: Academic Press 1983.

Rand, Gary M., ed. *Fundamentals of Aquatic Toxicology: Effects, Environmental Fate and Risk Assessment.* 2d ed. North Palm Beach: Taylor and Francis 1995.

Université Laval, Department of Geography. *Environmental Atlas of the St Lawrence.* Ottawa: Environment Canada 1992.

### Alien Invaders

Carlton, J.T., and J.B. Geller. "Ecological Roulette: The Global Transport of Nonindigenous Marine Organisms." *Science* 198 (1993): 394–6.

De Pinto, Joseph V., and Rajagopol Narayanan. "What Other Ecosystem Changes Have Zebra Mussels Caused in Lake Erie: Potential Bioavailability of PCBs." *Great Lakes Research Review* 3, no. 1 (1997): 1–8.

Dodson, Stanley I. *Ecology.* New York and Oxford: Oxford University Press 1998: 193–6.

Domske, Helen M, ed. "Great Lakes Exotic Species" (special issue). *Great Lakes Research Review* 3, no. 1 (1997).

Gavine, Kim. *Natural Invaders: Invasive Plants in Ontario.* Toronto: Federation of Ontario Naturalists 1996.

Gillis, P.L., and G.L. Mackie. "Impact of Zebra Mussels, *Dreissena polymorpha*, on Populations of Unionidae

(Bivalvia–Mussels) in Lake St Clair." *Canadian Journal of Zoology* 72 (1994): 1260–71.

Perlman, Dan L., and Glen Andelson. *Biodiversity: Exploring Values and Priorities in Conservation.* Malden: Blackwell Science 1997.

Reake-Kudla, Marjorie L., Don E. Wilson, and Edward O. Wilson, eds. *Biodiversity 2: Understanding and Protecting Our Biological Resources.* Washington: Joseph Henry Press 1997.



Ricciardi, Anthony, and Joseph B. Rasmussen. "Extinction Rates of North American Freshwater Fauna." *Conservation Biology* 13 (1999): 1220–2.

Ricciardi, Anthony, R.J. Neves, and J. B. Rasmussen. "Impending Extinctions of North American Freshwater Mussels (Unionoida) following the Zebra Mussel (*Dreissena polymorpha*) Invasion." *Journal of Animal Ecology* 67 (1998): 613–9.

– "Predicting the Identity and Impact of Future Biological Invaders: A Priority for Aquatic Resource Management." *Canadian Journal of Fisheries and Aquatic Sciences* 55 (1998): 1759–65.

Ricciardi, Anthony, Fred G. Whoriskey, and J.B. Rasmussen. "The Role of the Zebra Mussel (*Dreissena polymorpha*) in Structuring Macroinvertebrate Communities on Hard Substrata." *Canadian Journal of Fisheries and Aquatic Sciences* 54 (1997): 2596–2608.

– "Impact of the *Dreissena* Invasion on Native Unionid Bivalves in the Upper St Lawrence River." *Canadian Journal of Fisheries and Aquatic Sciences* 53 (1996): 1434–44.

Sea Grant Non-Indigenous Species Website: www.sgnis.org. (Related sites: www.ucc.uconn.edu/~wwwsgo/aen.html www.great-lakes.net/envt/exotic/exotic.html

**The Montreal Archipelago**

Bider, J.Roger, and S. Matte. *Atlas des amphibiens et des reptiles du Quebec.* Quebec, QC: Société d'histoire naturelle de la vallée du Saint-Laurent et ministère de l'Environnement et de la Faune du Québec, Direction de la faune et des habitats 1994.

Cyr, André, and J. Larivée. *Atlas saisonnier des oiseaux du Quebec.* Sherbrooke, QC: Les Presses de l'Université de Sherbrooke et la Société de loisir ornithologique de l'Estrie 1995.

Gauthier, J., and Y. Aubry (dirigé par). *Les oiseaux nicheurs du Quebec: Atlas des oiseaux nicheurs du Quebec méridional* (Breeding bird atlas of Quebec). Édité par l'Association québécoise des groupes ornithologues, la Société québécoise de protection des oiseaux, le Service canadien de la faune. Quebec, QC: Environnement Canada, Région du Quebec 1995.

Lanken, Dane. "Montreal Starts Cleaning up the St Lawrence." *Canadian Geographic* 110, no. 3 (1990): 28–35.

Marie-Victorin, Fr. *Flore Laurentienne.* 3d ed. Mise à jour par L. Brouillet, S.G. Hay, and I. Goulet in collaboration with M. Blondeau, J. Cayouette, and J. Labrecque. Montreal: Les Presses de l'Université de Montreal 1995.

Ministère de l'Environnement du Quebec. *L'environnement au Quebec. Un premier bilan rapport synthèse.* Quebec, QC 1988.

Ministère de l'Environnement et de la Faune. *Liste de la faune vertébrée du Quebec.* Quebec, QC 1995.

Scott, W.B., and E.J. Crossman. *Les poissons d'eau douce du Canada.* (Freshwater fishes of Canada). Service des pêches et des sciences de la mer, bulletin 184. Ottawa: Ministère de l'Environnement du Canada 1974.

Université Laval, Department of Geography. *Environmental Atlas of the St Lawrence.* Ottawa: Environment Canada 1992.

**Balancing Act**

Ceballos-Lascurain, Hector. *Tourism, Ecotourism and Protected Areas.* Gland, Switzerland: IUCN 1996.

Coaster's Association. *The Lower North Shore Tourism Guidebook*. Montreal: McGill University Printing Services 1999.

Hull, John S. "Market Segmentation and Ecotourism Development on the Lower North Shore of Quebec." In *Sustainable Tourism: A Geographical Perspective*, edited by C. Michael Hall and Alan Lew, 146–58. London: Longman 1998.

Lindberg, Kreg, and B. Mckercher. "Ecotourism: A Critical Overview." *Pacific Tourism Review* 1 (1997): 65–79.

Manning, Edward W. *Governance for Tourism: Coping with Tourism in Impacted Destinations*. Ottawa: Centre for a Sustainable Future, Consulting and Audit Canada 1998.

Rice, Michael. Project development assistant at the Mohawk Band Council and co-director of Mohawk Trail Tours, Box 405, Kahnawake,QC, Canada, J0L 1B0.

**Blocking the Flow of Rivers**

Delisle, C.E., and M.A. Bouchard, eds. *Managing the Effects of Hydroelectric Development*. Montreal: Canadian Society of Environmental Biologists 1990.

Hamie, R.S., ed. *Symposium on Small Hydropower and Fisheries*. Bethesda: American Fisheries Society 1985.

Joseph, Patrick. "The Battle of the Dams." *Smithsonian* 29, no. 11 (1998): 48–61.

Ward, J.V,. and J.A. Stanford, eds. *The Ecology of Regulated Streams*. New York: Plenum Press 1979.

**Great Rivers Meet**

Béland, Pierre. "The Beluga Whale of the St Lawrence River." *Scientific American* 274, no. 5 (1996): 74–81.

El-Sabh, M.I., and N. Silverberg. *Oceanography of a Large-Scale Estuarine System: The St Lawrence*. New York: Springer-Verlag 1990.

Environment Canada, Quebec Region, St Lawrence Centre. *State of the Environment Report on the St Lawrence River*. Vol. 1 & 2. Montreal: Éditions MultiMondes, Environment Canada, St Lawrence Centre 1996.

Environnement Canada, Région du Québec, Centre Saint-Laurent. *Rapport-synthèse sur l'état du Saint-Laurent*. Vol. 1 & 2. Montreal: Éditions MultiMondes, Environnement Canada, Centre Saint-Laurent 1996.

Jones, Linda, and Greg Shaw. "Saguenay Marine Park: Window on a Unique Environment." *Borealis* 3, no. 3 (1992): 28–33.

Ministry of Canadian Heritage and Ministère de l'Environnement et Faune du Quebec. *Crossroads of Life, Site of Exchange and Riches, Wellspring of Life: The Saguenay–St Lawrence Marine Park Management Plan*. Ottawa, ON, and Quebec, QC: Ministry of Canadian Heritage and Ministère de l'Environnement et Faune du Quebec 1995.

**Bi-National Citizen Action**

Adkin, Laurie E. *Politics of Sustainable Development: Citizens, Unions and the Corporations*. Montreal: Black Rose Books, 1998.

Manno, Jack. "Advocacy and Diplomacy in the Great Lakes: A Case History of Non-Governmental Organization Participation in Negotiating the Great Lakes Water Quality Agreement." *Buffalo Environmental Law Journal* 1, no. 1 (1993): 1–61.

Muldoon, Paul, and John Jackson. "Keeping the Zero in Zero Discharge: Phasing Out Persistent Toxic Substances in the Great Lakes Basin." *Alternatives* 20, no. 4 (1994): 14–20.

**Sustainable Development**

Keating, Michael. *Canada and the State of the Planet*. Toronto: Oxford University Press 1997.

Wackernagel, Mathis, and William Rees. *Our Ecological Footprint*. Gabriola Island, BC, and Philadelphia 1996.

World Commission on Environment and Development. *Our Common Future* (The "Brundtland Report"). Oxford and New York: World Commission on Environment and Development 1987.

World Conservation Union, United Nations Environment Programme and World Wide Fund for Nature. *Caring for the Earth: A Strategy for Sustainable Living*. Gland, Switzerland: World Conservation Union 1991.



### Frog Reflections

Alternatives. "Making Sense of the Ecosystem Approach: Lessons from the Great Lakes." *Alternatives* 20, no. 3 (1994).

– "Saving the Great Lakes" (special issue). *Alternatives* 13, no. 3 (1986).

Caldwell, Lynton K., ed. *Perspectives on Ecosystem Management for the Great Lakes: A Reader*. Albany, NY: State University of New York Press 1988.

Edwards, Clayton J., and Henry A. Regier, eds. *An Ecosystem Approach to the Integrity of the Great Lakes in Turbulent Times*. Ann Arbor, MI: Great Lakes Fishery Commission 1990.

MacKenzie, Susan Hill. *Integrated Resource Planning and Management: The Ecosystem Approach in the Great Lakes Basin*. Washington: Island Press 1996.

Weeks, W. William. *Beyond the Ark: Tools for an Ecosystem Approach to Conservation*. Washington: Island Press 1997.

### Lessons from a Woodpecker

Botts, Lee, and Bruce Krushelniki. *The Great Lakes: An Environmental Atlas and Resource Book*. 3d ed. Ottawa: Government of Canada; Chicago: United States Environmental Protection Agency 1995.

Hummel, Monte, ed. *Protecting Canada's Endangered Spaces: An Owner's Manual*. Toronto: Key Porter 1995.

– *Endangered Spaces: The Future for Canada's Wilderness*. Toronto: Key Porter 1989.

Killan, Gerald. *Protected Places: A History of Ontario's Provincial Parks System*. Toronto: Dundurn Press 1993.

Kraulis, J.A., and Kevin McNamee. *The National Parks of Canada*. Toronto: Key Porter 1994.

Marsh, John, and Bruce W. Hodgins, eds. *Changing Parks: The History, Future and Cultural Context of Parks and Heritage Landscapes*. Toronto: Natural Heritage/ Natural History 1998.

The Nature Conservancy (Great Lakes Program). *The Conservation of Biological Diversity in the Great Lakes Ecosystem: Issues and Opportunities*. Chicago: The Nature Conservancy 1994.

Wild Earth. "The Wildlands Project: Plotting a North American Wilderness Recovery Strategy" (special issues 1 and 2). *Wild Earth* (1992, Winter 1995/96).

Wright, Gerald R., ed. *National Parks and Protected Areas: Their Role in Environmental Protection*. Boston: Blackwell Science 1996.

### Why Rivers? Why Watersheds?

Notes

1 David Orr, *Ecological Literacy: Education and the Transition to a Post-Modern World* (State University of New York, 1992).

2 Otto Langer, "Summary Comments and Workshop Reflects," in *Urean Stream Protection: Restoration and Stewardship in the Pacific Northwest*. Workshop Proceedings, 10–12 March 1997.

3 James R.Karr, "Testimony to the Subcommittee on Energy and the Environment, House Interior and Insular Affairs Committee," U.S. Congress, 29 April 1992.

4 "At the Watershed of Environmental Management," *Water Connection*, newsletter of the New England Interstate Water Pollution Control Commission, vol. 10, no. 1 1993.

5 Lester R. Brown, et al. *Vital Signs 1993: The Trends That Are Shaping Our Future* (Worldwatch Institute 1993).

Books and Periodicals

Abbey, Edward. *One Life at a Time, Please*: Henry Holt 1988.

Benke, Arthur C. "A Perspective on America's Vanishing Streams." *Journal of the North American Benthological Society* 9, no. 1 (1990).

Brody, Jane E. "Water Based Animals Are Becoming Extinct Faster Than Others." *New York Times*, 21 April 1991.

Cairns, John, Jr. "Developing a Strategy for Protecting and Repairing Self-Maintaining Ecosystems." *Journal of Clean Technology and Environmental Science* 1991.

Cairns, John Jr, et al. *Restoration of Aquatic Ecosystems*. Washington: National Academy Press 1992.

Commission on Population Growth and the American Future. *Population and the American Future: The Report of the Commission on Population Growth and the American Future*. New York: Signet 1972.

Cone, Marla. "UV Light Killing Frogs." *Denver Post*, 1 March 1994: 1.

Council on Environmental Quality and the Department of State. *The Global 2000 Report to the President: Entering the Twenty-First Century*. Vols. 1 and 2. (Including *Global Future: Time to Act*). Charlottesville, VA: Blue Angel 1981.

Doppelt, Bob, et al. *Entering the Watershed: A New Approach to Save America's River Ecosystems*. Washington: Island Press 1993.

Master, Larry. "The Imperiled Status of North American Aquatic Animals." *Biodiversity Network News* (Arlington, VA) 3, no. 3 (1990).

Miller, R.R., J.D. Williams, and J.E. Williams. "Extinctions of North American Fishes during the Past Century." *Fisheries* 14 (1989): 22–38.

Yoffe, Emily. "Silence of the Frogs." *New York Times Magazine*, 13 December 1992: 36–39, 64–66, 76.

**Revelations**

Ashworth, William. *The Late Great Lakes*. New York: Alfred A. Knopf 1986.

Carson, Rachel. *Silent Spring*. New York: Fawcett Crest 1962.

Colborn, Theodora E., Alex Davidson, Sharon N. Green, R.A. (Tony) Hodge, C. Ian Jackson, and Richard A. Liroff. *Great Lakes, Great Legacy?* Washington, DC: The Conservation Foundation; Ottawa: The Institute for Research on Public Policy 1990.

Daly, Herman E., and John B. Cobb. *For the Common Good*. Boston: Beacon Press 1989.

Government of Canada. *The State of Canada's Environment*. Ottawa: Government of Canada Publications 1996.

Government of Canada. *Toxic Chemicals in the Great Lakes and Associated Effects, Synopsis*. Vols. 1 and 2. Ottawa: Environment Canada, Department of Fisheries and Oceans, Health and Welfare 1991.

International Joint Commission, 100 Ouellette Avenue, 8th Floor, Windsor, Ontario N9A 6T3, Tel. (519) 257-6700; Fax (519) 257-6740; e-mail: commission@windsor.ijc.org ; website: http://www.ijc.org .

Keating, Michael. *Canada and the State of the Planet*. Toronto: Oxford University Press 1997.

– *To the Last Drop: Canada and the World's Water Crisis*. Toronto: Macmillan 1986.

Leopold, Aldo. *A Sand County Almanac; With Other Essays on Conservation from Round River*. New York: Oxford University Press 1966.

McMichael, A.J. *Planetary Overload*. Cambridge: Cambridge University Press 1993.

National Wildlife Federation and Canadian Institute for Law and Policy. *A Prescription for Healthy Great Lakes.* Ann Arbor: National Wildlife Federation; and Toronto: Canadian Institute for Law and Policy 1991.

Postel, Sandra, Worldwatch Institute. *Last Oasis: Facing Water Scarcity.* New York: Norton 1992.

United Nations Department of Public Information. *Agenda 21: The United Nations Programme of Action from Rio.* New York: United Nations Department of Public Information 1993.

### Epilogue

Bolsenga, Stanley, and Charles Herdendorf. *Lake Erie and Lake St Clair Hankbook.* Detroit: Wayne State University Press 1993.

Cameron, Silver Donald. "But Nev'r a Drop to Drink." *Equinox 107* (1999): 30–9.

de Villiers, Marq. *Water.* Toronto: Stoddart 1999.

Ricciardi, Anthony, and Joseph B.Rasmussen. "Extinction Rates of North American Freshwater Fauna." *Conservation Biology* 13 (1999): 1220–2.

# Author Biographies

**Gregor Beck** is a wildlife biologist specializing in ecology, conservation, education, and watersheds. He has worked throughout Canada and for many years led projects for the Quebec-Labrador Foundation/Atlantic Center for the Environment and taught at Ryerson Polytechnic University and Seneca College. Gregor holds degrees from McGill University, St. Francis Xavier University, and the University of Guelph, and is author of *Watersheds: A Practical Handbook for Healthy Water* – a popular introduction to ecology and environmental issues. He is director of Conservation and Science for the Federation of Ontario Naturalists.

**Anne Bell** grew up beside the Avon River in Stratford, Ontario. Now living in Toronto, she is a director of the Wildlands League chapter of the Canadian Parks and Wilderness Society. She is completing her PhD in the Faculty of Environmental Studies at York University.

**J. Douglas Blakey** taught biology and environmental science at East Northumberland Secondary School before joining the faculty of Upper Canada College in Toronto in 1975, where he has been principal since 1991. He is a graduate of the Universities of Guelph and Western Ontario, and a board member of several educational organizations. A keen scuba diver, for many summers he also taught marine biology at the Huntsman Marine Science Centre in New Brunswick.

**Serge Bourdon** is a sculptor whose themes include social environment, nature, and the integration of art into architecture. He co-founded and directed the Atelier du Vieux Longueuil a multi-discipline artists' organization near Montreal, and was responsible for several national museum exhibits. In 1989 he settled along Quebec's Châteauguay River, where he operates a small nursery. He joined the Châteauguay River Rehabilitation Committee and helped found the Châteauguay Watershed Management Agency (SCABRIC), where he continues to serve as an active board member.

**Robert Brander** is a native of Michigan's Upper Peninsula. The son of an American lighthouse keeper on Lake Superior and grandson of a Canadian Lake Superior commercial fisherman, he retired in 1993 after over twenty years of service as an ecologist with the U.S. Forest Service and National Park Service. Dr Brander lives on a hill in Wisconsin overlooking Lake Superior.

**Dominique Brief**'s professional interests lie in natural resource conservation. Her recent focus is tourism's potential to alleviate local pressures on wildlife and their habitat. Her past activities span the globe, from the Baffin Region of Northern Canada to Jamaica in the Caribbean, examining ecotourism's potential as a conservation strategy. She consults for public, private, and non-governmental organizations and is director of sustainable tourism projects at Alliance for Environmental Management (AEM).

**Louise Champoux** has been a wildlife toxicologist with the Canadian Wildlife Service of Environment Canada since 1990, where she evaluates exposure and effects of environmental contaminants on wildlife. She obtained her MSc in biology and aquatic toxicology from the University of Montreal in 1986 and worked for the Inland Waters Directorate and the St Lawrence Centre on the contamination of St Lawrence River sediments and waters.



**Bruce Conn** is professor of biology and dean of the School of Mathematical and Natural Sciences at Berry College in Rome, Georgia. He has conducted extensive research on the biota of the St Lawrence River. His writings include an award-winning textbook and more than one hundred other scientific publications.

**Kevin Coyle** is an attorney who served for ten years with the U.S. Department of the Interior's National Park Service and Bureau of Outdoor Recreation. He co-founded and was first president of American Land Resource Association. While he was president of American Rivers, the organization protected more than twenty thousand miles of river and five million acres of riverside land. He served as vice-president, programs, River Network; in 1996, he became president of the National Environmental Education and Training Foundation, Inc. in Washington.

**Brad Cundiff** is a freelance writer with a long-standing interest in environmental issues. His work has appeared in many of Canada's leading environment and ecology publications, and he has written two books on exploring Southern Ontario. Brad is a director of the Wildlands League.

**Jerry Valen DeMarco** is a staff lawyer and registered professional planner with the Sierra Legal Defence Fund. He has served as a board member of the Wildlands League and the Canadian Parks and Wilderness Society. His recent publications have appeared in *Protecting Canada's Endangered Spaces*, *Canada's River Heritage*, and *Canadian Issues in Environmental Ethics*.

**Jean-Luc DesGranges** holds a PhD in terrestrial ecology from McGill University and joined the Canadian Wildlife Service in 1977 as a research scientist, studying colonial freshwater birds including great blue herons. In the 1980s and early 1990s, he studied the impacts of acid rain, contaminants, and pesticides on wildlife. He now conducts multidisciplinary research aimed at developing multi-species indicators of biodiversity for forest management and conservation in the St Lawrence region.

**Thomas A. Edsall** is a fishery research biologist with the U.S. Geological Survey, Biological Resources Division, Great Lakes Science Center in Ann Arbor, Michigan. He is currently chief of the Western Basin Ecosystem Branch, with responsibility for fishery and aquatic research in Lakes Superior and Michigan, and terrestrial and aquatic research in the five Great Lakes National Lakeshore "Cluster" Parks. He has authored more than eighty scientific papers and technical reports and is actively pursuing ecological research on Great Lakes fish and benthic invertebrates.

**Peter Ewins** is director of World Wildlife Fund Canada's Species Program. He previously worked with the Canadian Wildlife Service, studying ecotoxicology of fish-eating birds on the Great Lakes. His doctorate from Oxford University was on the effects of the North Sea oil industry on seabirds in Shetland. He is married with two young daughters and lives in Toronto.

**Louis-Gilles Françoeur** holds degrees in philosophy and political science (BA, Université Saint-Paul; MA, Université de Montreal) and began his journalism career in Ottawa in the 1960s with *Le Droit*, where he also worked in broadcasting and with government. He has reported for *Le Devoir* (Montreal) since the early 1970s, specializing in environment since 1982. He is one of Canada's most experienced environmental reporters, and the recipient of numerous journalism awards.

**Stephen Gates** of Coaticook, Quebec, has been personally and professionally involved in river conservation for over twenty years as land planner, consultant to the U.S. Environmental Protection Agency, college professor, and lawyer in private practice. He is now president of the Grey Owl Nature Trust, which supports projects that protect and restore Canada's wilderness, waters, and wildlife.

**Elliott Gimble** directed community-based river and watershed conservation programs at the Quebec-Labrador Foundation in Ipswich, Massachusetts, from 1989 to 1997. Previously, he worked at the Smithsonian Institution coordinating public programs for a travelling exhibition on tropical rainforests. He holds degrees from Dartmouth College and Yale University's School of Forestry and Environmental Studies. He currently works for the Jewish Community Relations Council in Boston.

**Hallett J. (Bud) Harris** received his PhD in zoology from Iowa State University; from 1978–1986 he coordinated research in Green Bay, Lake Michigan, with University of Wisconsin's Sea Grant Institute. From 1988 to 1992 he worked with the EPA Great Lakes Program Office determining the fate of PCBs and other toxins. He is the Herbert Fisk Johnson Professor of Environmental Studies and Chair of Natural and Applied Sciences, University of Wisconsin–Green Bay.



**John Hull** received his PhD from McGill University in 1999. His dissertation examined the potential for tourism to promote sustainable economic development on the Lower North Shore of Quebec. Since the mid 1980s he has worked on conservation and development projects for the Quebec-Labrador Foundation and has also been a consultant for the World Bank, UNESCO, and the Commission on Environmental Cooperation.

**Gail Jackson**, BSc, was educated in southern Ontario and moved to the northern part of the province in the late 1970s. She works with Parks Canada, and has served within government as a geologist, biologist, and environmental planner for over twenty years. With her passion for the natural environment, she spends much of her time on the Superior coast fishing with her husband, Mike, and skipping rocks and watching falcons with her two young sons, Christopher and Nicolas.

**John Jackson** is a citizen activist who has worked full time with citizens' groups on environmental contamination and waste management issues for the past twenty years. He has been a board member of Great Lakes United for fifteen years, including six years as president.

**Michael Keating** is an international environment writer and consultant, with a special knowledge of water issues. He is author of six books, including *To the Last Drop, Canada and the World's Water Crisis, The Earth Summit's Agenda for Change*, a report on environment and development, and *Canada and the State of the Planet.*

**Val Klump** is a senior scientist and biogeochemist at the University of Wisconsin–Milwaukee Great Lakes Water Institute. His research focuses on nutrient and carbon cycling in lakes, and his laboratory has become a leader in the development and use of undersea robotics. He holds a BS degree in zoology from Duke University, a degree in law from Georgetown University, and a PhD in chemical oceanography from the University of North Carolina.

**Louise Knox** has been co-ordinator of the Hamilton Harbour Remedial Action Plan since 1994. An employee of Environment Canada since 1989, she facilitated public involvement in RAPs and served as head, Policy and Planning, in Toronto. Earlier, she consulted in public affairs and worked with the Canadian Mental Health Association. She holds a BA in French literature from the University of Western Ontario and studied French and French-Canadian literature at the University of Ottawa.

**Gail Krantzberg** has worked for Ontario's Environment Ministry since 1988 as aquatic ecotoxicologist, coordinator for the Great Lakes Remedial Action Plan Program, and is currently Great Lakes strategic policy analyst. Her research includes bioassessment, ecosystem restoration, and stormwater, and she also interprets technical information to the public. She is president of the International Association of Great Lakes Research and on the faculty of the University of Toronto; she has published over eighty articles in refereed and other journals.

**Peter Lavigne** is an attorney, activist, writer, and educator. He served as executive director of Westport River Watershed Alliance and Merrimack River Watershed Council, co-founded the Coalition for Buzzards Bay and New England Coastal Campaign, and worked for American Rivers in Quebec, New England, and New York. He directed River Network's national River Leadership Program (1992–1996). Pete is president of Watershed Consultants, an international strategy and policy business in Portland, Oregon, and teaches graduate water law and policy at Portland State University.

**Michel Letendre** is a biologist who received his degree from the University of Sherbrooke in 1977. He worked for the Society for the Development of James Bay and the Ministry of Transport of Quebec. Since 1987 he has worked as a marine life specialist with Quebec's Ministry of the Environment and Fauna and is a knowledgeable and passionate researcher on the aquatic habitats of the Montreal and Monteregie Archipelago.

**Bruce Litteljohn** has been associated with Upper Canada College in Toronto as teacher, administrator, or consultant since 1965, and was founding director of the Norval Outdoor School. He was a director of the Wildlands League for twenty-three years, being particularly active in removing logging from Quetico Provincial Park. His photographs and writings are widely published, and he is co-author or co-editor of several books including *Islands of Hope: Ontario's Parks and Wilderness* and *Superior: The Haunted Shore.* He lives in Bracebridge, Ontario, where he is on the Natural Heritage Committee of the Muskoka Heritage Foundation. He holds degrees from McMaster University, the University of British Columbia, and University of Toronto.

**Nadia Ménard**, a Quebec-born biologist, has worked at the Saguenay–St Lawrence Marine Park since 1992 and holds BSc and MSc degrees from Laval University. She has studied the feeding environment of St Lawrence estuary whales, has worked in different areas between the Great Lakes Area and Mingan Islands National Park, and is deeply committed to conservation and education. She lives with her husband, Ian, in picturesque Tadoussac, at the confluence of the Saguenay and St Lawrence Rivers.

**Jeff Miller** moved to Canada from the U.S. in 1957 to enjoy wilderness. For over fifty years he has been intimately connected with Ontario's Algonquin Provincial Park, canoe tripping and guiding, and later painting, writing, and speaking on behalf of wilderness protection. He was a founding director and president of the( Algonquin) Wildlands League and co-creator of "Look-See-Paint," a video and paint-kit program that promotes art to save nature.

**Phillip Norton** lives with his wife and two children in Châteauguay, Quebec, and works in the photography department of the *Montreal Gazette.* Prior to that, he made his living as weekly newspaper editor, photographer, and freelance magazine writer, always with a focus on environmental issues such as acid rain, nuclear waste, and river clean-ups. He serves as volunteer director on the board of the Châteauguay Watershed Management Agency.

**Jean Rodrigue** has been a biologist with the Canadian Wildlife Service of Environment Canada since 1990. He obtained his Master's degree from McGill University in 1994. He has been involved in different projects including contamination in great blue herons in the St Lawrence River, mercury poisoning in ospreys in hydroelectric reservoirs in James Bay, and frog deformities in agricultural habitats.

**Alec Ross** is a writer and journalist and author of *Coke Stop in Emo: Adventures of a Long-Distance Paddler*. The book describes an eight-thousand kilometre solo canoe trip from Montreal to Vancouver that he completed between 1987 and 1989. He lives in Kingston, Ontario, with his wife, Vicki Westgate, and their daughter, Madeleine.

**Scot Stewart** is a science teacher and naturalist who has lived on Lake Superior in Marquette, Michigan, since 1969. He hitchhiked around Lake Superior in 1976 and has since written about and photographed the lake for magazines and books. He continues to hike, paddle, and explore with his wife, Corinne Rockow, and their two daughters, to create photograph-music-story programs. He is a former director of Great Lakes United.

**Rae Tyson**, an independent journalist from Clifton, Virginia, is co-founder and president of the 1,200-member Society of Environmental Journalists in the United States. An environmental reporter and editor at *USA Today* for over a decade, he began his journalism career covering Love Canal at Niagara Falls, New York, in the 1970s. A former high school and college science teacher, he authored a book on environmental risks affecting children, published by Random House in 1995.

**Fred Whoriskey** earned his BSc at Brown University and his PhD from Laval University. He has worked at the Woods Hole Oceanographic Institution, as a professor at McGill University, as deputy scientific coordinator for the Great Whale Hydro Project Scientific Support Office, and is presently vice-president of Research and Environment for the Atlantic Salmon Federation.

# Acknowledgments



We would like to express our sincere thanks to the many individuals, groups, and financial supporters who helped to make this project possible. We have attempted here to name the key people and organizations, and we apologize to those whose names we have accidentally omitted. The *Voices for the Watershed* project is vast in scope and has been long in the making; the number of persons and partners who have helped along the way reflects this history.

First, we thank the Wildlands League (a chapter of the Canadian Parks and Wilderness Society) for endorsing this educational project, and the National Undersea Research Center (University of Connecticut, Avery Point Campus) for providing core financial support. In total, fifteen sponsors and friends funded the research, writing, and production of this book, and we recognize these foundations and organizations formally on the Sponsors page vi.

We thank the staff and directors of the Wildlands League for their advice, support, and help with fundraising, in particular, Tim Gray (Executive Director) and Jane Truemner, as well as Anne Bell, Brad Cundiff, Jerry DeMarco, Deborah Freeman, and Bruce Mackenzie. At National Undersea Research Center, we thank Ivar Babb, Doug Lee, Richard Cooper, Peter Sheifele; and the U.S. National Oceanic and Atmospheric Administration's Aquanaut Program. Sincere thanks also to Captain James Seiler and First Mate George Gunther and crew of the Research Vessel *Edwin Link* (Harbor Branch Oceanographic Institute).

We acknowledge with thanks the hard work and patience of the many contributing authors and photographers who have made this international project a reality; many donated their work, and for that we are particularly appreciative. Due to budgetary and production constraints, we were not able to include all writings that were submitted for consideration; we thank these authors also, and extend our sincere regrets. We extend special thanks to Doug Lee, John Lee, and Michael Keating, who worked with us as part of the book's creative team; and, to Monte Hummel (president, World Wildlife Fund Canada) for writing the Foreword.

Our friend and colleague John Lee (V. John Lee Communication Graphics Inc.) was involved in the project from the beginning, and deserves special mention. We are extremely grateful to John for helping to develop the concept of the book, for his involvement in project planning, and finally for designing this handsome anthology. We extend our thanks to William Gilpin for indexing the book, and to Julie O'Brien (Federation of Ontario Naturalists) for researching the conservation organizations' addresses. We would also like to express our appreciation to McGill-Queen's University Press for publishing this work, and in particular to Maureen Garvie, copy editor, Joan McGilvray, coordinating editor, Susanne McAdam, production design manager, and the staff at the press. We also thank Eric Lienberger (University of British Columbia) for preparing the map.

Many individuals assisted the editors with research, field work, logistics, and fundraising for the project. We express our appreciation to: Christopher Baines, Paola Battison; Angèle Blasutti (Partnership for Public Lands); Guy Bouchard; Bracebridge Public Library staff; the late Jack Christie (Ontario Ministry of Natural Resources); Sue Doerr (World Wildlife Fund Canada); Rob Cromwell; Gwen and the late Wayland Drew; Tom Farnquist and Mark Warner (Great Lakes Shipwreck Historical Society, Michigan); Jacques Hébert, Nadia Ménard and Marc Pagé (Parks Canada); Christiane Hudon and Nicole Lavigne (St Lawrence Centre, Environment Canada); Kathleen Hurd; Gail Krantzberg (Ontario Ministry of the Environment); Josh Laughren (World Wildlife Fund Canada); Henry Lickers (Mohawk Council of Akwesasne); James Loates (cartographer); the late Madeleine Mardall, and Barbara Shettler; Anne May and Joel Cooper; Rose McConnell; Ken McMillan (McQuest); Jim Murphy (Ontario Ministry of Natural Resources); Mike and Carolyn O'Connor; Barb Olds and Betsy Landry (Finger Lakes); Dale Park (Upper Canada College) ; Bruce Petrachuk; Charles Schafer (Geological Survey of Canada); David Sergeant (Fisheries and Oceans Canada); Chip Weseloh (Canadian Wildlife Service); Dirk Van Wijk (Owl Rafting); Michael Williams (Walpole Island First Nation); Susan Woods; Ian Wyndlow and Hugue Vienne.

Anthony Ricciardi (Dalhousie University) reviewed "Alien Invaders," and his suggestions and additions significantly improved this chapter on exotic species. Dominique Brief and John Hull extend their thanks to the residents of the Quebec Lower North Shore for their support and assistance in producing the guide books, and to the Quebec-Labrador Foundation and Quebec Ministry of Education for financial support. They also thank Michael Rice (Mohawk Band Council of Kahnawake and Mohawk Trail Tours) for providing information about the Kahnawake Native Reserve and its tourism ventures.

Finally, we thank our families and our friends for their support and encouragement throughout the project.

# Index